ELECTRONIC CHIPS & SYSTEMS DESIGN LANGUAGES

Electronic Chips & Systems Design Languages

Edited by

Jean Mermet

Directeur de recherches au CNRS, ECSI, France

KLUWER ACADEMIC PUBLISHERS
BOSTON / DORDRECHT / LONDON

A C.I.P. Catalogue record for this book is available from the Library of Congress.

ISBN 978-1-4419-4884-7

Published by Kluwer Academic Publishers,
P.O. Box 17, 3300 AA Dordrecht, The Netherlands.

Sold and distributed in North, Central and South America
by Kluwer Academic Publishers,
101 Philip Drive, Norwell, MA 02061, U.S.A.

In all other countries, sold and distributed
by Kluwer Academic Publishers,
P.O. Box 322, 3300 AH Dordrecht, The Netherlands.

Printed on acid-free paper

TABLE OF CONTENTS

Contributors

Albenge M.F. - Allara A. - Alonso A. - Amann H.P. - Ashenden P. - Aubert V. - Baidas Z.A. - Bakalar K. - Barros E. - Bauer M. - Berry G. - Bjureus P. - Boegli A. - Bombana M. - Borrione D. - Brown A.D. - Budde R. - Calvez J.P. - Carpenter A. - Cerny E. - Chenard E. - Christen E. - Clouté F. - Comai S. - Contensou J.N. - Cook F. - Corvino D. - Dabrowski J. - De Araujo C. - Droegehorn O. - Ecker W. - Ellervee P. - Epicoco I. - Esteve D. - Favard Y. - Fernandez L.S. - Ferrandi F. - Fummi F. - Garcia-Sabiro S. - Geisselhardt W. - Georgelin P. - Groba A. - Haase J. - Harcourt E. - Heller D. - Hemani A. - Houzet D. - Huemmer H.D. - Jantsch A. - Josko B. – Kumar S. - Kumar R. - Lavagno L. - Martin D. - Martinez N. - Messer N. - Moser V. - Muller F. - Nebel W. - Nicolae A.F. - Oberg J. - O'Nils M. - Pampagnin P. - Pasquier O. - Pellandini F. - Pickin S. - Ploeger P.G. - Pons P. - Pulka A. - Radetzki M. - Sander I. - Schlör R. - Schwarz P. - Sciuto D. - Sentovich E. - Svantesson B. - Sylla K.H. -Williams A.C. - Wilsey P. - Zinn A.

Preface

The Forum on Design Languages (FDL) is the best opportunity to exchange experiences and to learn about new efforts and trends in the application of languages and notations and their associated design methods and tools in electronic systems design.

FDL98 and FDL99, the first occurrences of this Forum, have brought together three events under the same roof, namely the HDL workshop, including the VHDL User's Forum in Europe (VUFE), the Workshop on Virtual Components and Design Re-use (VCDR) and the Workshop on System Specification and Design Languages (SSDL). VUFE is the European VHDL User's Group yearly meeting on VHDL related topics and standardization efforts. It was founded at the VLSI'89 Conference in Munich by the project that was going to become ECSI and by the IFIP WG 10.2 (now 10.5). VCDR started in 1995 as the Workshop on Libraries, Component Modeling, and Quality Assurance, and has merged with the Design Re-use workshop. It is the European event on all aspects related to the creation, the use and re-use of virtual components in electronic design. SSDL started in 1996 as SLDL to address the need to develop a general industry-wide consensus on the key problems of specifying and designing embedded distributed or on-a-chip electronic systems.

As you will notice the "Electronic Chips and Systems Design Languages" selection integrates HDL, VUFE, VCDR and SSDL topics as different facets of the challenge faced by designers to-day.
- The practical and effective use of VHDL and other hardware description languages in the areas of system design, design validation and test, design synthesis, and component modeling.
- The HDL and SLDL international standardization process, including insights of the future languages and formal notations.
- The reality of design re-use practice, including components modeling aspects, and report on industrial experience from European projects or from VSI-alliance's working groups .

FDL98 and FDL99 have taken place in prestigious academic institutions, before the start of the new academic year, namely the Ecole Federale Polytechnique de Lausanne in 1998 and the Ecole Normale Supérieure de Lyon in 1999.

I hope that this selection will be the first of a long series, for the best profit of students, designers and managers in the domain of electronic systems design.

Jean Mermet

VHDL EXTENSIONS

ANALOG AND MIXED-SIGNAL MODELLING AND SIMULATION

Introduction by Alain Vachoux, (chair VHDL-AMS IEEE standardisation group)

Since its inception in 1998, the Forum on Design Language has dedicated a special attention to new developments around hardware description languages (HDLs) that bring new capabilities to analog and mixed-signal modelling and simulation. One main reason was, and still is, the emergence of new languages, such as VHDL-AMS or Verilog-AMS, that are in the process to be supported in many leading EDA environments. These languages basically extend the event-driven modelling and simulation capabilities that already exist in VHDL or Verilog with new notations and semantics to describe and to simulate analog and mixed-signal systems. It should be noted that the extensions are not limited to electrical systems so modelling and simulation of, say, mechanical or hydraulic systems, and combinations of these (so-called multi-disciplinary systems) is possible.

This chapter presents revised and augmented versions of a selection of four papers that have been presented in FDL98 and FDL99.

The first paper, *"Efficient modelling of analog and mixed A/D systems using PWL technique"* by Jerzy Dabrowski and Andrzej Pulka, proposes a way to use "digital" VHDL for the modelling and the simulation of analog and mixed-signal systems. The base of the approach is the use of a piecewise linear modelling technique that basically makes a continuous model event-driven. The approach is well suited for functional analog models that may be mixed with digital behavior. Analog models with tight feedback loops are efficiently handled and, although iterations may be needed during simulation, the simulation time and the accuracy of the results still compare favorably with what can be obtained with an analog simulator.

The second paper, *"Library development using the VHDL 1076.1 language"* by Ernst Christen and Ken Bakalar, presents guidelines to ensure interoperability between VHDL-AMS, the informal name for VHDL 1076.1, descriptions. The paper first proposes a standard infrastructure for modelling multi-disciplinary systems in the form of a VHDL-AMS package including declarations of useful physical constants, tolerance names and natures related to physical disciplines. The paper then suggests modelling techniques to build a library of electrical (SPICE) models that would make use of the new package.

J. Mermet (ed.), Electronic Chips & Systems Design Languages, 2–3.
© 2001 *Kluwer Academic Publishers.*

The third paper, *"Behavioral modelling of complex heterogeneous systems"* by Peter Schwarz and Joachim Haase, discusses the issues related to the support of various modelling methods such as network/conservative models, block/signal-flow models, or bond graph models, in analog hardware description languages in order to describe and simulate multi-disciplinary systems such as microsystems. The paper presents a canonical mixed-signal/mixed-discipline description method that may be formulated in currently available HDLs.

The fourth and last paper, *"VHDL-AMS, an unified language to describe multi-domain, mixed-signal designs"* by Véronique Aubert and Serge Garcia-Sabiro, defines a subset of the VHDL-AMS language and modelling guidelines to support mixed-discipline mechatronic systems. The paper reports on experiences in the European project TOOLSYS that partly intended to address modelling and simulation issues in automotive applications. The paper also presents application examples in the electromagnetic and in the mechanical domains.

It is my hope that the four papers in this chapter will provide useful insights in the capabilities of emerging standard analog and mixed-signal hardware description languages. The improving quality of EDA environments that support these languages is likely to push designers to use these capabilities more and more in system design. Forthcoming FDL events will be the preferred places to report on these experiences.

Alain Vachoux

April 2000

Library Development Using the VHDL-AMS Language

Ernst Christen, Kenneth Bakalar

Analogy, Inc.   *Mentor Graphics Corporation*

Key words: VHDL-AMS, 1076.1, model library

Abstract: The VHDL 1076.1 standard for mixed analog/digital systems was approved by the IEEE in 1999. However, the design goal of portability of models and designs cannot be met without additional effort. We suggest some guidelines for the use of the extended language and propose the development and standardization of a package that supports multi-disciplinary modeling and a library of basic electrical models. Both will aid in the deployment and acceptance of VHDL-AMS tools in industry and academia. We offer a proposal for joint development of the standard with the Verilog-AMS constituency, to provide some interoperability between the two languages.

1. INTRODUCTION

IEEE Std. 1076.1-1999 [1] extends the VHDL language defined by IEEE Std. 1076-1993 [2] to support the description and simulation of analog and mixed analog/digital systems. The extended language defined by the two standards has informally been called VHDL-AMS.

The VHDL-AMS language allows a user to write analog, digital and mixed-signal models. However, to achieve portability of models between different tools, a set of guidelines and companion standards is necessary, similar to the IEEE 1164-1993 companion standard [3] to the VHDL 1076 language. Our goal is to identify the areas where companion standards are useful or even essential, and to demonstrate how a number of modeling problems that appear to require auxiliary standards can instead be addressed with the existing facilities of VHDL-AMS. Although our focus is on analog modeling, we assume that the reader is familiar with the VHDL language.

5

J. Mermet (ed.), Electronic Chips & Systems Design Languages, 5–16.
© 2001 *Kluwer Academic Publishers.*

2.   OVERVIEW OF VHDL-AMS

VHDL-AMS is a superset of the VHDL 1076-1993 language. Any legal VHDL 1076-1993 description is also legal in VHDL-AMS and gives the same simulation results, with the exception of conflicts caused by the introduction of new reserved words. The language supports the hierarchical description and simulation of digital (i.e. discrete), analog (i.e. continuous), and mixed-signal (i.e. mixed analog/digital) systems. The analog portion of the model consists of lumped elements that can be described by algebraic and ordinary differential equations (jointly called differential-algebraic equations, or DAEs). Models can have connections with conservative (i.e. satisfying, in electrical systems, KCL/KVL) and signal-flow semantics. Support is provided for electrical and non-electrical disciplines (or energy domains) at levels of abstraction ranging from the system level to the SPICE-like circuit level. The language supports flexible and efficient interactions between the event-driven, digital computation engine and the continuous, analog computation engine. Finally, support for small-signal AC and noise simulation in the frequency domain is provided.

We now describe the facilities of the VHDL-AMS language related to modeling of analog devices.

The language represents an unknown in the DAEs by a *quantity*. Scalar quantities must be of a floating-point type. *Free quantities* support signal-flow modeling and may also represent values like the power dissipated in a resistor or the charge in a capacitor. *Branch quantities*, which are declared with reference to two *terminals*, support the modeling of conservative systems. There are two kinds of branch quantities. *Across quantities* represent the potential difference between two terminals—in electrical systems, the voltage between two nodes. *Through quantities* represent the flow between two terminals—in electrical systems, the current in a branch between two terminals. Finally, *source quantities* describe small-signal frequency domain sources and noise sources.

Terminals support conservative connections. Any number of connected terminals form a node at which KCL (or its generalization to non-electrical disciplines) is enforced. Terminals have a *nature* that defines the discipline of the terminal—electrical, thermal, rotational, etc. Only terminals with like natures can be connected together. The nature also defines the types of the across quantities and through quantities incident to a terminal of that nature and the reference terminal for the discipline—in electrical systems, ground.

The equations describing the behavior of an analog model are written using *simultaneous statements*. Simultaneous statements support two modeling styles and two mechanisms to specify piecewise defined behavior, i.e. behavior that is different in different regions of operation. The most basic member of this class of statements is the simple simultaneous statement; its simplified syntax is

expression == expression ;

When a solution of the DAEs has been found, the value of the expression on the left-hand side of the == sign is close to the value of the expression on the right-hand side.

In general, the DAEs must be solved using numerical methods. Suitable algorithms, developed over the past thirty years, find solutions subject to user-defined *tolerances*. Unfortunately, each algorithm uses tolerance parameters slightly differently, so it is not possible to provide a universal scheme based on a vector of real values. The language addresses this issue by defining the concept of *tolerance groups*: each quantity and each equation belongs to a tolerance group. The model writer associates a tolerance group with each quantity and equation; then, before simulation begins the model user associates the numerical tolerance parameters appropriate for his simulator with each tolerance group.

3.   LIBRARY DEVELOPMENT

A library of models that is intended for a large audience must be portable between language implementations and interoperable with user-defined models in each environment. We observe that many analog modelers are not sufficiently familiar with language-based modeling in general or VHDL-AMS in particular. These considerations motivate our discussion.

In the following we define an infrastructure supporting the development of a portable library, demonstrate how the language can be used to solve a number of modeling problems, propose a minimum content of a portable library of electrical models, and outline a process to define such a library.

3.1   Infrastructure for Modeling Multi-Disciplinary Systems

Just as the VHDL 1076 language defines no built-in logic types, the VHDL 1076.1 language defines no built-in natures. The declarative apparatus of VHDL-AMS must be used to provide these and other appropriate extensions in the form of VHDL-AMS packages. The following tasks are prerequisites to the development of a portable library:
- The definition of a collection of standard natures that support the modeling of both electrical and non-electrical systems.
- The definition of the units used to represent physical quantities.
- The definition of a suite of standard tolerance groups.

- Guidelines for the use of tolerance groups.
- Declarations of physical and mathematical constants and mathematical functions that are commonly used in writing analog models.

3.1.1 Natures

Physical systems obeying conservation semantics are generally modeled using the concepts of across and through variables [5, 6]. The sum of all through variables incident to a node is zero. The sum of all across variables in any closed loop is zero. Each physical discipline has different across and through variables, as shown in Table 1.

Table 1. Selected Physical Disciplines

Discipline	Across Variable (unit)	Through Variable (unit)
electrical	electromotive force (Volt)	current (Ampere)
magnetic	magnetomotive force (Ampere*turn)	magnetic flux (Weber)
thermal	temperature (Kelvin)	heat flow (Watt)
translational	velocity (meter/second)	force (Newton)
rotational	angular velocity (radians/second)	torque (Newton*meter)
fluidic	pressure (Pascal)	flow rate (meter^3/second)

Sometimes, alternative across variables are used for translational and rotational disciplines: displacement and angle.

To emphasize the correspondence between the physical world and the concepts of the VHDL-AMS language, and to facilitate international acceptance of the library, we propose the following conventions:

- The portable library is based on the SI units [7] as shown in Table 1.
- The name of a nature is the name of the physical discipline it represents.
- The name of the across or through type of the nature is derived from the SI name of the corresponding across or through variable [7].
- The name of the reference terminal of the nature is derived from the name of the nature.

The VHDL attribute mechanism can be used to support the annotation of quantities with units. An attribute of type string associates a unit name with each across and through type. A waveform viewer can use this information to annotate a graph of simulation results with the proper units.

As an example, the following declarations define subtypes, natures and units for electrical systems:

```
attribute symbol: STRING;
subtype current is REAL;
attribute symbol of current: subtype is "A";
subtype emf is REAL;
attribute symbol of emf: subtype is "V";
nature electrical is emf across current through
        electrical_ref reference;
```

3.1.2 Tolerances

Tolerance groups allow a model writer to group quantities whose value over time are to be determined with the same accuracy. A tolerance group is represented by a tolerance code of type string attributed to the subtype of a quantity. More than one subtype may share the same tolerance code.

What criteria should be used to select tolerance groups for quantities? Numerical mathematicians recommend that tolerances should be selected "to accurately reflect the scale of the problem" [9]. We conclude that quantities whose values over time will have a similar range should by default belong to the same tolerance group. For many applications this means that, for example, all voltages are in one tolerance group and all currents in another. A suitable default is to define a tolerance group for each subtype used in a nature definition. With this recommendation the subtype declarations of the previous example should be changed to:

```
subtype current is REAL tolerance "default_current";
subtype emf is REAL tolerance "default_emf";
```

The defaults can be overwritten in a model, which may be necessary, for example, when a design consists of low-voltage circuitry controlling a high-voltage application. Models can even be written with parameterized tolerance groups, as shown in the following entity declaration of a resistor.

```
entity resistor is
      generic (r: REAL;
             vtol: STRING := voltage'Tolerance;
             itol: STRING := current'Tolerance);
      port (terminal p, m: electrical
             tolerance vtol across itol through);
end entity resistor;
```

The default values of vtol and itol are the tolerance codes of the subtypes voltage and current.

A particular simulator must provide some mechanism to associate appropriate tolerance parameters with a tolerance group. The meaning of the tolerance parameters will depend on the algorithms used by the simulator. The values of the parameters chosen for a particular tolerance group will depend on the question a user wants to answer about the design; the user can trade off accuracy against simulation speed by adjusting the parameters.

A simulator may support the specification of tolerance parameters in a variety of ways. For example, it may provide suitable defaults for the tolerance parameters of each tolerance group; it may allow the tolerance parameters for several tolerance groups to be entered together; it may support the derivation of the tolerance parameters of one tolerance group from those of another.

3.1.3 Mathematical and Physical Constants, Mathematical Functions

The behavior of analog models often depends on mathematical and physical constants and involves mathematical functions such as exponential, logarithmic and trigonometric functions. The recently standardized package ieee.math_real [4] defines a collection of mathematical constants and functions that should be sufficient for all but the most advanced modeling needs. A large number of physical constants have been defined in [8]; we propose to include a subset of this collection in the infrastructure of a portable library.

3.1.4 A Package for Multi-Disciplinary Modeling

To facilitate the development of portable libraries of models, the infrastructure described in this section should be defined in a VHDL package and made available as an IEEE standard. This package has the same importance for VHDL-AMS as package std_logic_1164 [3] has for VHDL 1076. The package should include as a minimum:
- Natures for the physical disciplines listed in Table 1.
- For each nature, the definition of a scalar nature, an unconstrained one-dimensional array nature, and the subtypes necessary for the definition of these natures including tolerance codes and units.
- A collection of physical constants, including at least the charge of an electron, the speed of light, Boltzmann's constant, Planck's constant, the permittivity and permeability of vacuum, the conversion between Kelvin and degrees Celsius, and selected material constants.
- Possibly other definitions, for example two-dimensional array natures, additional subtypes like charge, additional natures like a rotational nature with angle as across variable.

3.2 Modeling Techniques

We now describe techniques for solving a number of common problems encountered when modeling electronic devices. The solutions are probably not obvious to the reader of the language definition document.

3.2.1 SPICE Model Parameters

Semiconductor device models have two kinds of parameters:
- Physics based parameters whose values are dependent on the manufacturing process. These parameters are typically shared by many instances of a device model.

– Geometry based parameters whose values are different for different instances of the device model.

To simplify the creation of circuit descriptions (i.e. netlists) the SPICE simulator [10] defines the concept of a *model* that contains values for all physics based parameters of a device. To avoid confusion with the use of the term "model" to refer to a set of device equations or to their implementation in a hardware description language, we will call such SPICE models *parameter sets* in what follows. As an example, the SPICE statement

.MODEL dd D IS = 1e-15 N = 1.5

defines a parameter set named dd for a diode. The definition specifies values for the parameters IS and N; all other parameters of the parameter set will take on their default values as defined for the diode. The parameter set can then be used by many diode instances, by specifying its name as a parameter of a diode instance, together with instance-specific parameters:

D1 2 3 dd 2.5

where D1 is the name of the diode instance, 2 and 3 are the nodes connected to its anode and cathode, and 2.5 is its area.

A parameter set can be represented in VHDL-AMS by a constant of a record type together with a constructor function that returns a value of that type. Each parameter of the parameter set is a named element of the record type and also a parameter of the constructor function. The default value of each parameter of the function is the default value of the corresponding parameter of the device. The entity declaration for the device to be parameterized has a generic whose type is the record type and whose default is defined by a call to the function.

As an example, consider a (simplified) parameter set for a diode model. The corresponding declarations are:

```
type DiodeModel is record
    is0:  REAL;                 -- saturation current
    n:    REAL;                 -- emission coefficient
    kf:   REAL;                 -- flicker noise coefficient
    af:   REAL;                 -- flicker noise exponent
end record;
function DiodeModelValue (
    is0:  REAL := 1.0e-14;      -- saturation current
    n:    REAL := 1.0;          -- emission coefficient
    kf:   REAL := 0.0;          -- flicker noise coefficient
    af:   REAL := 1.0)          -- flicker noise exponent
                        return DiodeModel is
begin
    return DiodeModel'(is0, n, kf, af);
end function DiodeModelValue;
```

```
entity Diode is
    generic (model: DiodeModel := DiodeModelValue;
             area: REAL := 1.0) -- instance parameter
    port (terminal anode, cathode: electrical);
end entity Diode;
```

With these definitions it is possible to declare a constant whose value is the same as the value of the parameter set defined by the SPICE .MODEL statement that appears above:

```
constant dd: DiodeModel :=
        DiodeModelValue (is0 => 1.0e-15, n => 1.5);
```

and then to use this constant as an actual in the association list of the generic map of an instance of the diode:

```
d1: Diode generic map (model => dd, area => 2.5)
           port map (anode => t2, cathode => t3);
```

Additional constructor functions could also be defined; for example one that takes an existing value of the type as a starting point and overwrites specified elements. A constructor function may check whether the parameters are within the bounds supported by the model, a necessity for production quality models. For maximum flexibility the definitions should be placed in a package, so they can be used in different design entities. For complete SPICE compatibility, the constant itself can be declared in a package, so it, too, can be accessed in different design entities.

3.2.2   Ambient Temperature

The behavior of most physical devices depends on operating temperature. In general, the operating temperature of a device depends on the ambient temperature and on how the device is operated (self-heating). Most first level models, including all SPICE models, ignore the effects of self-heating and only consider the dependency of the model equations on the ambient temperature. Furthermore, the ambient temperature is considered constant. For models of electronic devices this approximation is valid if the power dissipation in the device and in other devices in close proximity is small. To allow a user to investigate the behavior of a circuit as a function of temperature, simulators typically support the concept of a temperature sweep, where the ambient temperature is set to a sequence of user-specified values and a suite of simulations is run at each temperature.

The same effect can be accomplished in VHDL-AMS by defining the ambient temperature as a deferred constant in a package and making it visible where necessary. A more general solution that supports different ambient temperatures in different instances defines the ambient temperature as a generic of a design entity whose initial value is the ambient temperature from the package. This solution makes it possible to use instances of the same design entity in parts of a design that are at different temperatures, for example one part at room temperature, another in a refrigerated environment.

3.2.3 Mutually Coupled Inductors

Two inductors are mutually coupled if their magnetic lines of force link together. The effect of this coupling is that any change of the current through one inductor induces a voltage in the other inductor. Transformers are an example of the application of mutually coupled inductors: the inductances of its windings are mutually coupled.

The SPICE approach to describing two mutually coupled inductors uses two instances of an inductor model and one instance of a special coupling element. The coupling element is not a stand-alone model; it can only be instantiated together with the two inductors. A different approach is required in VHDL-AMS since there is no way to describe dependencies of this kind.

Considering that a two-winding transformer consists of two mutually coupled inductors, we propose to implement mutual inductance as a transformer model. However, it is an easy task to extend the model to N windings, any two of which are mutually coupled. Assuming that a type real_matrix has been defined and that the "*" operator has been overloaded for multiplying a real matrix and a real vector, we can implement the N-winding transformer as follows:

```
entity xformer is
    generic (ml: real_matrix);
    port (terminal p, m: electrical_vector);
end entity xformer;
architecture lossless of xformer is
    quantity v across i through p to m;
begin
    v == ml * real_vector(i'dot);
end architecture lossless;
```

where ml is a symmetric square matrix with diagonal elements representing the inductance values L_i and an off-diagonal element at position (i,j) representing the mutual inductance value M_{ij} between inductors L_i and L_j: $M_{ij} = k_{ij} \cdot \sqrt{L_i \cdot L_j}$. k_{ij} is the coupling factor used by SPICE.

3.2.4 Noise Modeling

Noise is a summary term for unwanted fluctuations in the value of a continuous waveform. Noise can have a number of different origins; mathematically it is treated as a random process. Its properties can be described by time series in the time domain and by noise spectra in the frequency domain. Our discussion will focus on noise modeling in the frequency domain, for time domain noise modeling see [11].

Noise modeling in the frequency domain is supported in VHDL-AMS by the concept of noise source quantities. The value of a noise source quantity during a noise calculation is specified by its power spectral density; the value is defined to be zero for all other calculations. The expression defining the power spectral density can depend on the simulation frequency, and the spectrum can be bias dependent.

Thermal Noise is caused by the random motion of charge carriers due to thermal excitation. Thermal noise can be modeled as white noise, i.e. with a constant power spectral density. As an example, consider the model of a noisy resistor. Its thermal noise current is described by $i_{th}^2 = 4kT\Delta f / R$ where k is Boltzmann's constant, T is the temperature in Kelvin, R is the resistance value, and f is the frequency. In the following architecture the noise current is represented by the noise source quantity i_thermal. The expression defining its power spectral density corresponds directly to the definition of the thermal noise current. The noise source quantity is simply added to the simple simultaneous statement that defines the behavior of the resistor. See section 3.1.2 for the entity declaration of the resistor.

```
architecture noisy of resistor is
    quantity v across i through p to m;
    quantity i_thermal: current noise
            4.0 * physical_K * ambient_temperature / r;
begin
    i == v / r + i_thermal;
end architecture noisy;
```

Flicker Noise, also called 1/f noise, is associated with the generation and recombination of charge carriers. It is always associated with a current i and is described by $i_f^2 = K_f |i|^{A_f} \Delta f / f$ where K_f is the flicker noise coefficient, A_f is the flicker noise exponent, and f is the frequency. In a diode, flicker noise depends on the diode current and is therefore bias dependent.

The flicker noise relationship is readily expressed in a VHDL-AMS diode model (see section 3.2.1 for the corresponding entity declaration):

```
architecture simple of Diode is
    quantity v across id through anode to cathode;
    quantity i_flicker: current noise
            model.kf * abs(id) ** model.af / FREQUENCY;
    constant vt: REAL :=
        physical_K * ambient_temperature / physical_Q;
begin
    id == area * model.is0 * (exp(v/(model.n*vt)) - 1.0) +
                        i_flicker;
end architecture simple;
```

where vt is the thermal voltage. The language definition guarantees that id has the value at the quiescent point when the expression specifying the power spectral density is evaluated. As before, the noise source quantity is added to the simple simultaneous statement defining the device behavior.

3.3 A Library of Basic Electrical Models

While the definition of a package for multi-disciplinary modeling greatly facilitates the portability of models written using these definitions, it only partially addresses the issue of making designs portable. To illustrate the problem, each SPICE vendor has implemented his proprietary modifications to the SPICE models. As a consequence, the same SPICE netlist may be read into several different SPICE simulators, but simulations will yield different results because the models are different. A similar situation may exist with simulators supporting VHDL-AMS, since each VHDL-AMS vendor will have to provide a library of basic electrical models to its customers.

The problem can be addressed by the development and standardization of a library of basic electrical models. Besides aiding product introductions, the library increases the portability of designs and improves the acceptance of the language. A good starting point for a standardized library of basic electrical models is the collection of SPICE models [10], which includes:
– ideal models for linear devices such as resistor, capacitor, inductor, controlled sources, and voltage and current sources
– semiconductor models such as diode and transistors (MOS and bipolar)
– some additional models, for example models for transmission lines

The library definition should include the scope of the library and complete semantics for each model in the library, including interface definitions and definitions of the behavior of each model. The implementation should consist of an entity declaration for each model that specifies its interface, and a package containing type declarations, definitions of constructor functions, component declarations, and other infrastructure. To make the library flexible to use, the approaches of section 3.2 to some of the modeling issues will be useful.

3.4 Standardization of Packages

Since several vendors and universities are currently developing products supporting the VHDL-AMS standard, we cannot delay much longer the definition of a package supporting multi-disciplinary modeling and the definition of a library of basic electrical models. Recognizing that Verilog-AMS tools will face similar issues, and in an attempt to provide some interoperability between VHDL-AMS and Verilog-AMS, we propose to

have the package and the library developed and standardized in a language-independent way, in cooperation with the Verilog-AMS constituency. From a procedural perspective there are two possible approaches:
- Form a study group that investigates these issues in more detail and eventually turns into a working group.
- Make this effort part of PAR 1076.5. This may be difficult or even impossible if the effort includes, as proposed, Verilog-AMS.

It should be possible to rapidly complete definitions for both the package and the library using either approach.

4.   CONCLUSIONS

We have given a brief overview of the capabilities of the VHDL-AMS language. We have shown how the language can be used to address a number of modeling problems, many of them related to approaches familiar to SPICE users. We have proposed the development and standardization of a package supporting multi-disciplinary modeling and of a library of basic electrical models. We believe that both the package and the library are essential for portability of models and designs. A joint development of the package and the library with the Verilog-AMS constituency could provide some interoperability between the two languages.

REFERENCES

[1] IEEE Standard VHDL Analog and Mixed Signal Extensions, IEEE Std. 1076.1-1999
[2] IEEE Standard VHDL Language Reference Manual, ANSI/IEEE Std 1076-1993
[3] IEEE Standard Multivalue Logic System for VHDL Model Interoperability
    (Std_logic_1164), IEEE Std 1164-1993
[4] IEEE Standard VHDL Language Mathematical Packages (MATH_REAL and
    MATH_COMPLEX), IEEE Std 1076.2-1997
[5] F.E. Cellier: *Continuous System Modeling*, Springer-Verlag, 1991
[6] J.W. Lewis: *Modeling Engineering Systems*, HighText Publications, 1994
[7] *The International System of Units (SI)*, NIST Special Publ. 330, 1991 Edition, and
    Guide for the Use of the International System of Units (SI), NIST Special Publ. 811, 1995
    Edition, available at http://physics.nist.gov
[8] Fundamental Physical Constants, in *Journal of Research of the National Bureau
    of Standards*, vol. 92, no. 2, March-April 1987, available at http://physics.nist.gov/cuu
[9] K.E. Brenan, S.L. Campbell, L.R. Petzold: *Numerical Solution of Initial-Value
    Problems in Differential-Algebraic Equations*, North-Holland, 1989
[10] SPICE versions 2G and 3F, available through the Software Distribution Office, Cory
    Hall, University of California, Berkeley, CA 94720
[11] N.J. Kasdin: *Discrete Simulation of Colored Noise and Stochastic Processes
    and 1/f**a Power Law Noise Generation*, Proc. IEEE, vol. 83, no. 5, May 1995, pp.
    802-827

BEHAVIORAL MODELING OF COMPLEX HETEROGENEOUS MICROSYSTEMS

Peter Schwarz and Joachim Haase

Fraunhofer Institut für Integrierte Schaltungen (IIS) Erlangen
Außenstelle EAS Dresden
Zeunerstraße 38
D-01069 Dresden, Germany
{schwarz, haase}@eas.iis.fhg.de

A general mathematical description of the terminal behavior of subsystems is discussed. It allows to combine different modeling methods, e. g. network models, block models, and bondgraphs together. The resulting equations can be solved with a general network analysis program. This approach is effectively applied in modeling electrical as well as non-electrical subsystems. Extensions to digital and time-discrete subsystems are possible. Therefore, modeling the terminal behavior of subsystems is the basis for simulation of heterogeneous systems. Extensions to digital and time-discrete subsystems are possible. Therefore, modeling the terminal behavior of subsystems is the basis for simulation of complex microsystems.

1 INTRODUCTION

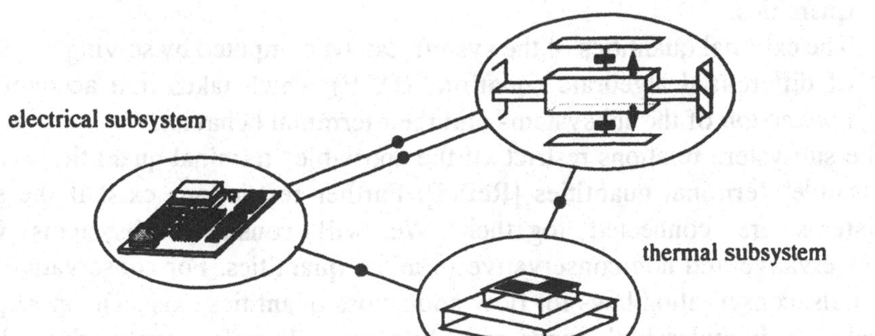

Figure 1. Subsystems of a microsystem

Microsystem design is a highly interdisciplinary area and, therefore, different CAD methods and tools have to be used together. Simulation plays an important role in the design of microsystems. We have developed an approach for the unified treatment of subsystems in different physical domains. Mechanical, thermal, electrical, or hydraulic parts of microsystems (Figure 1.)

17

J. Mermet (ed.), Electronic Chips & Systems Design Languages, 17–30.

18

and their interactions may be modelled on the basis of a generalized multi-pole concept and the definition of the terminal behavior. The model description may be implemented in standardized hardware description languages (HDLs).

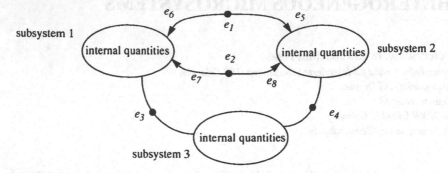

Figure 2. System partitioned into subsystems with external quantities e_i

At first, we will restrict the considerations on continuous systems and will discuss extensions later. The approach is based on the following ideas (see Figure 2.):

- The whole system is divided into subsystems. Quantities are divided into *external* quantities, which connect the subsystems, and *internal* quantities of the subsystems. The external quantities at a subsystem build its *terminal quantities*.
- The behavior of the subsystems depends only on their terminal quantities (and their initial conditions, too). Therefore, a mathematical modeling approach should be used which formulates the *relations* between terminal quantities.
- The external quantities of the system can be computed by solving a system of differential algebraic equations (DAE) which takes into account the *connection* of the subsystems and their terminal behavior.

The subsystem relations restrict all the „possible" terminal quantities to „admissible" terminal quantities [Rei85]. Further restrictions exist if the subsystems are connected together. We will consider subsystems with conservative and non-conservative terminal quantities. For conservative terminals, conservation laws for flow and across quantities exist. The most general case includes both kinds of subsystems. Therefore, network analysis programs like Saber, ELDO or Spectre should be used. They establish the system of differential equations using the terminal behavior descriptions of the subsystems. These descriptions may be formulated by using special description languages like Mast, HDL-A, SpectreHDL or in future VHDL-AMS [VHDL]. With block diagram-based simulators like Matrix$_X$ or Matlab it is difficult to describe conservative systems and to handle the algebraic loops which are typical for DAE systems.

The information exchange between subsystems is done by the quantities at their terminals. This paper gives an unified approach to the modeling of such systems which combines well-known modeling methods and summarizes experiences in using this approach. It can be shown that various modeling methods may be interpreted as special cases of a general approach based on behavioral terminal descriptions.

2   DESCRIPTION OF THE TERMINAL BEHAVIOR

We start with a classification of the pins which can be divided into conservative and non-conservative terminals:
- Conservative terminals carry a through quantity (or flow quantity) i and an across quantity v (current and voltage, resp., in the electrical domain).
- Non-conservative terminals carry only a quantity a (which may be interpreted as an across quantity, too).

This distinction is very useful to include two important classes of systems: *Kirchhoffian networks* and signal-flow based *block diagrams*. The time-continues, real-valued quantities are functions $x: R \rightarrow R$. The quantities of all terminals of a subsystem are combined in the vectors i, v and a (see Figure 3.). The dimension of i and v is equal to the number of conservative terminals, the dimension of a is equal to the number of non-conservative terminals. A connection point of terminals in a system is called a *node*. The simulator has to guarantee that the computed through and across quantities at a node are admissible terminal quantities. In addition the sum of the through quantities at a node has to be zero.

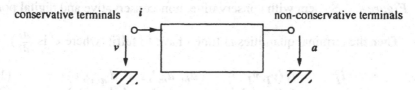

conservative terminals i non-conservative terminals

Figure 3.   Terminals of a subsystem

There is no general approach to formulate the model equations. But in many cases further refinements and special assumptions give valuable hints to describe the terminal behavior in an easy way. E. g., terminal quantities $i = (i_1, i_2)$, $v = (v_1, v_2)$ and $a = (a_{in}, a_{out})$ may be strictly partitioned into dependent and independent quantities $d = (i_1, v_2, a_{out})$ resp. $u = (i_2, v_1, a_{in})$. The formulation of the terminal behavior may require *additional (free) quantities s* which may be related to internal states, internal node voltages, or auxiliary quantities [CHS96]. To consider delay effects, e. g. in systems with transmission lines, it is necessary to use not only actual quantity values $x(t)$ but also the former values, in general the whole history of a quantity $x(t)$. This is abbreviated as $x|t$ (quantity x until time t). Introducing additional (free) quantities s, further equations are added to describe the relations of the terminal quantities d and u and the free quantities s. The dependent quantities at time t can be described as a function F_1, the additional equations between u and s are given by a function F_2:

$$d(t) = F_1(u|t, s|t, p)$$
$$0 = F_2(u|t, s|t, p)$$

p is a vector of *parameters* of the subsystem. The *terminal behavior* can always be described in this form. To go into more detail, we will restrict our considerations to models which do not use explicitly the history of quantities.

The approach can also be extended to subsystems with digital and time-discrete ports (see Figure 4.). These ports are divided into input ports with signals d_{in} and output ports with signals d_{out}. Therefore, the digital port signals have to be included into the vector of dependent and independent signals $d = (i_1, v_2, a_{out}, d_{out})$ resp. $u = (i_2, v_1, a_{in}, d_{in})$.

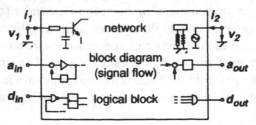

Figure 4. System with conservative, non-conservative and digital ports

Then the terminal quantities at time t have to fulfil (where x' is $\frac{dx}{dt}$)

$$i_1 = f_1(v_1, v_1', i_2, i_2', a_{in}, a_{in}', s, s', d_{in}, p, t) \qquad (1)$$
$$v_2 = f_2(v_1, v_1', i_2, i_2', a_{in}, a_{in}', s, s', d_{in}, p, t) \qquad (2)$$
$$a_{out} = f_3(v_1, v_1', i_2, i_2', a_{in}, a_{in}', s, s', d_{in}, p, t) \qquad (3)$$
$$0 = f_4(v_1, v_1', i_2, i_2', a_{in}, a_{in}', s, s', d_{in}, p, t) \qquad (4)$$
$$d_{out} = f_5(v_1, v_1', i_2, i_2', a_{in}, a_{in}', s, s', d_{in}, p, t) \qquad (5)$$

It is possible (with a few restrictions) to transform this set of equations into a HDL in a systematic way. We focused our consideration on VHDL-AMS, Mast, and HDL-A. But other languages like Verilog-A will offer similar features. As an other example for a multi-disciplinary language Modelica [Mod] should be mentioned. It is an object-oriented modeling language which covers also modeling approaches like networks, block diagrams, and bond graphs. Multi-disciplinary libraries will be developed and may be used in microsystem modeling and simulation. In the following the skeleton of a VHDL-AMS model is given:

```
ENTITY entity_name IS                    -- description of the component interface
    GENERIC  ( p : ... );                -- parameters
    PORT     (TERMINAL t1: ... ; QUANTITY aout : ...  ; SIGNAL  dout : ... );
END ENTITY entity_name;

ARCHITECTURE name OF entity_name IS
    QUANTITY  v1 ACROSS  i1  THROUGH  t1;     -- declaration of terminal quantities
    QUANTITY s : ... ;                        -- additional free quanties,
    CONSTANT ...;                             -- constants, ...
BEGIN
    simultaneous statements in accordance to (1, 2, 3) for
        the dependent terminal through quantities i_1,
        the dependent terminal across quantities v_2, and
        the non-conservative output quantities a_{out}
    simultaneous statements for bringing f_4 to zero (4)
    concurrent statements to determine d_{out} with respect to (5)
END ARCHITECTURE name;
```

Main advantages of this canonical description of the terminal behavior that will be discussed in the next paragraphs are

- Applicability of standardized or widely-used *hardware description languages* (MAST, HDL-A, SpectreHDL, VHDL-AMS).
- Applicability of powerful *multi-level, mixed-mode simulators* which are part of many industrial design systems.
- The canonical form of the terminal description is a good basis for multi-disciplinary *component libraries.*
- Basic components (e.g. beams and plates for mechanical micro system modeling) may be combined to much more *complicated system models.* Re-use is supported in this way very effectively.
- The unified mathematical description is *independent* of simulators, but may be transformed easily into the behavioral languages of different simulators.
- Simulators like Saber, ELDO, and Spectre support behavioral description of subsystems as well as circuit descriptions. Therefore, a pure behavioral modelling approach is not necessary. Instead of, *mixed behavioral-structural modeling* may be applied.
- *Simulator coupling* may be interpreted as a special kind of evaluation of the terminal behavior.

3   DIFFERENT DESCRIPTION METHODS

3.1   Network Models and Control Systems

Some commonly used subsystems like resistances, capacitances, inductances etc. are predefined elements in network analysis programs. They can be combined with subsystems described by (1) ... (4). On the other hand f_1 and f_2 can describe subsystems of different physical domains. In this case the flow and across quantities at the terminals have to be interpreted in a domain-specific way, e. g. as currents and voltages in electrical subsystems, as forces and displacements in mechanical subsystems etc.

Using analogies between different physical domains and the construction of generalized networks is a classical approach for modeling non-electrical systems [KoB61], [Len71], [Res61], [GeD97]. It is also possible that conservative terminals of subsystems belong to different physical domains. This allows the consideration of interactions inside a subsystem between different physical domains in a very natural way. For instance selfheating of electrical components (Figure 5.) is such an effect [Lei97]. These interactions are especially important in modeling microsystems.

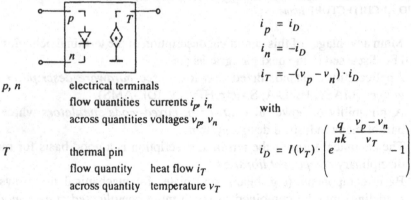

$$i_p = i_D$$
$$i_n = -i_D$$
$$i_T = -(v_p - v_n) \cdot i_D$$

p, n	electrical terminals
	flow quantities currents i_p, i_n
	across quantities voltages v_p, v_n

with

T	thermal pin
	flow quantity heat flow i_T
	across quantity temperature v_T

$$i_D = I(v_T) \cdot \left(e^{\frac{q}{nk} \cdot \frac{v_p - v_n}{v_T}} - 1 \right)$$

Figure 5.   Diode with selfheating

Control (sub)systems can be modeled if only non-conservative terminals are used.

3.2   Bondgraphs

Bondgraphs [KaR74], [Cel91], [Tho90] proved as a very popular modeling approach in some disciplines, especially if an energy flow is the basic physical phenomenon. They consist of bonds (like elements) and junctions (connection of elements). It is possible to transform them into a network or into a block diagram. An example shows how to transform a bond into a block diagram (Figure 6.).

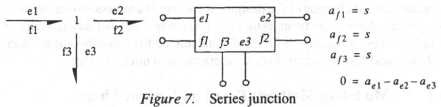

Case 1: $a_e = R \cdot a_f$

Case 2: $a_f = \frac{1}{R} \cdot a_e$

bond_R (caus_F)

bond_R (caus_E)

Figure 6. R element

In modeling bondgraphs, a pair of non-conservative pins carries a flow and an effort quantity (see also [CHS96]). It has to be distinguished whether flow results from effort or vice versa. Depending on the causality, flow or effort are non-conservative input and output terminals, respectively. Modeling a series junction is shown in Figure 7. Parallel junctions can be modeled in a similar way. The introduction of an additional free quantity s allows a formulation without special causality requirements to the incoming bonds.

$$a_{f1} = s$$
$$a_{f2} = s$$
$$a_{f3} = s$$
$$0 = a_{e1} - a_{e2} - a_{e3}$$

Figure 7. Series junction

4   EVALUATION OF THE TERMINAL BEHAVIOR

A second aspect of modeling heterogeneous microsystems is the usage of systematic methods for the computation of the terminal behavior.

4.1   Simulator Coupling

The functions $f_1 \dots f_4$ (see equations (1) - (4)) which describe the terminal behavior of a subsystem can be computed with another simulator. Simulator coupling realizes this idea and may be illustrated by an example (Figure 8.).

An acceleration sensor can be simulated with a FEM-program like AN-SYS. By the program ANSYS the displacement of the seismic mass is computed as a function of the acceleration a. In a system simulation, e. g. with the program Saber, a model of the sensor must be used. Inside the Saber simulation the evaluation of the terminal behavior of the sensor corresponding to f_1 $\dots f_4$ can be done now by ANSYS

24

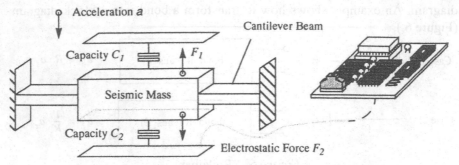

Figure 8. System with capacitive acceleration sensor

The system simulator (Saber) proposes a new set of independent terminal quantities (external quantities) and the FEM simulator (ANSYS) calculates the "response" terminal quantities, the dependent quantities. The system simulator is responsible for the correction of the terminal quantities to fulfill the conservation laws. Thus for the system simulation the same model as in the design process of the subsystem can be used [Kle95], [Ecc96], [WCS97]. The consequences of changes of the design parameters of a subsystem on the overall system behavior can be investigated in this way. We used this approach in coupling circuit simulators with FEM simulators in the simulation of mechanical-electrical and electrical-thermal interactions in microsystems.

4.2   Modeling Methods Based on System Theory

In some cases it is possible to derive approximation functions from data points of terminal quantities in the time or frequency domain. These data points can be given as a result of measurements or simulations of the subsystem to be modeled e.g. with a FEM-simulator. For linear dynamic and nonlinear static systems there exists a general way to establish models for simulation. However, the disadvantage of this method is that it doesn't give a deep insight into the function of a subsystem and the influence of design parameters to the system's behavior.

Linear dynamic subsystems: the dependent output quantities of linear dynamic systems are given by the convolution of input quantities and the impulse response of the model. This convolution can be carried out by different methods. Good results have been achieved with recursive convolution [Ngu94], [VoH95]. In this case the impulse response has to be developed as a sum of exponential functions.

Nonlinear static subsystems: many approximation methods are described in the textbooks (see [BoD87], e.g.). As a very powerful approximation method for static nonlinear subsystems radial basis functions can be used [JaC92], [Par97].

Nonlinear dynamic subsystems: in linear system theory, a lot of other approximation methods was developed which may be used in modeling. But for the most general case of modeling nonlinear dynamic systems only some specialized approaches exist and, therefore, this topic is an open question (see e. g. [Ise92, p. 223]). First approaches to derive nonlinear models from FEM descriptions are restricted to special classes of problems [SAW97].

All these algorithms may be embedded into VHDL-AMS model code via a C interface. The way how to do this will depend on the realization of the Foreign Language Interface in different simulators. An idea how it could work is given by the procedural interface of HDL-A.

4.3   Symbolic Analysis

Using the above mentioned analogies, programs for symbolic analysis (e.g. Analog Insydes) may be applied for the *formulation* of the terminal behavior [GiS91], [HeS95]. The method may help also to *reduce* an analytical description of a subsystem. This way gives a deeper impression how special parameters influence the terminal behavior as numerically based approximation methods can do. Especially for linear dynamic subsystems this method may be very effective but the applicability to nonlinear circuits is under investigation [Bor97]. Many problems like handling a very large number of system equations or terms in these equations have to be solved.

4.4   Physically based Modeling

Physically based modeling starts with formulating the basic relations between system variables, e. g. setting-up partial differential equations (PDE) of distributed systems or formulation of ordinary differential-algebraic equations of multi-body systems. We will focus on some modeling aspects important in microsystem technology. The PDE's used in modeling microsystem components are usually solved with FEM solvers. Other methods are based on Finite Differences (FDM), Boundary Element Methods (BEM), ... We will try to use modeling results of FEM solutions in the context of behavioral modeling, terminal description, and network simulators. Field-oriented modeling approaches in relation to network models are investigated also in [VoW92], [Wac95].

One systematic way to get behavioral models is now to derive functions f of subsystems similar to the way to establish finite elements for FEM solvers. The idea behind the Finite Element Method is to determine an (energy) functional of a system as the sum of the functionals of the subsystems [ZiT89],

[Sch84], [Sch93]. We use energy as such a functional. Then the energy of a subsystem can be determined with the help of some quantities at specified points at the border of the subsystem. These points are called *nodes*. The objective of the Finite Element Analysis is to determine the node quantities so that the energy of the system is *minimized*.

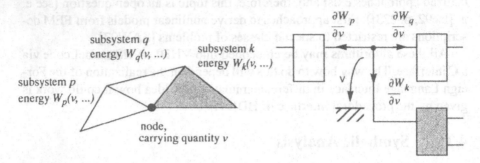

Figure 9. Part of a system and KCL for finite element model (principle)

Figure 9. shows a detail of a system. The energies of the subsystems p, q and k depend on the quantity v at the node. The energy W of the system is given by

$$W = \ldots + W_p(v, \ldots) + W_q(v, \ldots) + W_k(v, \ldots) + \ldots$$

The energy W is a minimum if all the derivatives to all node quantities are equal to zero. That means, the derivative of W to v has also to be zero. Because only W_p, W_q and W_k depend on v it follows

$$\frac{\partial W}{\partial v} = \frac{\partial W_p}{\partial v} + \frac{\partial W_q}{\partial v} + \frac{\partial W_k}{\partial v} = 0$$

This equation can be interpreted as a generalized Kirchhoff's Current Law (KCL) at the node. Thus the same description used in FEM-programs can (in principle) be used in network analysis programs. The „FEM-network" has to be composed of blocks which represent the subsystems. We use as flow quantities through the terminals the derivatives of the energy of the subsystem with respect to the across quantities at the terminals.

Example: Spar element

The deformation energy of a spar element (Figure 10.) depends on the displacement v in direction of the spar

$$W_s = \frac{1}{2} \cdot \iiint_V \sigma^T \varepsilon dV = \frac{E \cdot A}{2} \cdot \int_0^L \left(\frac{dv}{dl}\right)^2 dl \quad \text{with} \quad \varepsilon = \frac{dv}{dl} \quad \text{and} \quad \sigma = E \cdot \varepsilon$$

(V volume of the spar element, E modulus of elasticity).

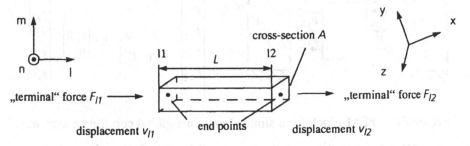

Figure 10. Spar element in a local coordinate system (l, m, n)

It follows under the assumption $v(l) = v_{l1} + \dfrac{v_{l2} - v_{l1}}{L} \cdot l$ (with $v_{l1} = v(l=0)$ and $v_{l2} = v(l=L)$)

$$\frac{\partial W_s}{\partial v_{l1}} = \frac{E \cdot A}{L}(v_{l1} - v_{l2}) \quad \text{and} \quad \frac{\partial W_s}{\partial v_{l2}} = \frac{E \cdot A}{L}(v_{l2} - v_{l1})$$

Thus, for the „FEM-block" that describes the behavior in the locale coordinate system (l, m, n) we set

$$F_{l1} = \frac{\partial W_s}{\partial v_{l1}} \quad \text{and} \quad F_{l2} = \frac{\partial W_s}{\partial v_{l2}}$$

That means the terminal behavior of a spar element parallel to the l-axes is given by

$$\begin{bmatrix} F_{l1} \\ F_{l2} \end{bmatrix} = \begin{bmatrix} \dfrac{E \cdot A}{L} & -\dfrac{E \cdot A}{L} \\ -\dfrac{E \cdot A}{L} & \dfrac{E \cdot A}{L} \end{bmatrix} \cdot \begin{bmatrix} v_{l1} \\ v_{l2} \end{bmatrix} = \bar{K} \cdot \begin{bmatrix} v_{l1} \\ v_{l2} \end{bmatrix}$$

The stiffness matrix K in a local (l, m, n)-coordinate system can be derived from $\bar{K}$.

Modeling complex micromechanical systems, composed of spar elements in different positions and other basic elements (e. g. beams), the formulation of the terminal equations in a *global* coordinate system (x, y, z) is more convenient (see Figure 11.). F_{di} and v_{di} are the flow and across quantities at conservative terminals of the „FEM-block". v_{di} can be interpreted as displacement in direction d (equal to x, y or z) at terminal i. C is a matrix to transform values from a local (l, m, n)-system to a global (x, y, z)-system. If (x_1, y_1, z_1) and (x_2, y_2, z_2) are the end points of a spar element the interesting matrix elements of C are

$$c_{11} = \frac{x_2 - x_1}{L} \quad , \quad c_{12} = \frac{y_2 - y_1}{L} \quad \text{and} \quad c_{13} = \frac{z_2 - z_1}{L}$$

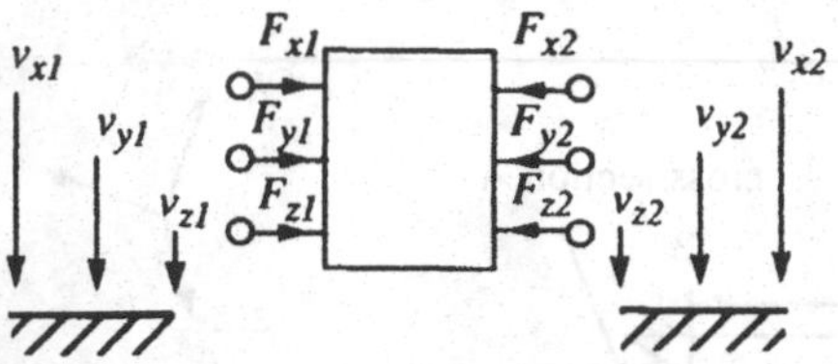

$$
\begin{bmatrix} F_{x1} \\ F_{y1} \\ F_{z1} \\ F_{x2} \\ F_{y2} \\ F_{z2} \end{bmatrix} = \begin{bmatrix} C^T & 0 \\ 0 & C^T \end{bmatrix} \cdot K \cdot \begin{bmatrix} C & 0 \\ 0 & C \end{bmatrix} \cdot \begin{bmatrix} v_{x1} \\ v_{y1} \\ v_{z1} \\ v_{x2} \\ v_{y2} \\ v_{z2} \end{bmatrix}
$$

Figure 11. FEM-block for a simple beam in a global coordinate system

In a network consisting of behavioral models of such spar elements external forces have to be taken into consideration as flow sources. Additional conditions concerning the displacements at special nodes are realized by across sources connecting the corresponding node and ground.

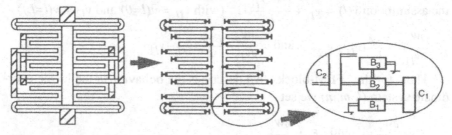

Figure 12. Principle of an acceleration sensor and its decomposition into beam and connection elements

An application of this approach [Neu98], [LoN98] is described in Figure 12. The principle of an acceleration sensor and its decomposition into beam elements and some connection elements is shown. The behavioral models of the beam elements extends the equations for the simple beam as shown in Figure 11. by dynamic effects and takes into consideration the rotational movements.

FEM simulators may be used also to formulate the description of (discretized) distributed subsystems. In some cases, the internal model description may be *exported* and may be reformulated as the terminal behavioral description. Grid meshing, discretization, and the field-oriented model descriptions may then be used in the system simulation, too. The FEM simulator ANSYS supports this behavioral modeling method by the „substructuring mode“ for exporting differential equations with an adjustable degree of freedom. We are developing a tool for the generation of behavioral models based on these exported model description.

For geometrical arrangements with a relatively simple structure it is possible to generate models for system simulation without using sophisticated FEM or FDM simulators. We developed a tool for the generation of static thermal models (Figure 13.) of chips and other microsystems [Wün98] (see also [SzR98). The generated models may be simulated together with other models in a mixed-domain simulation.

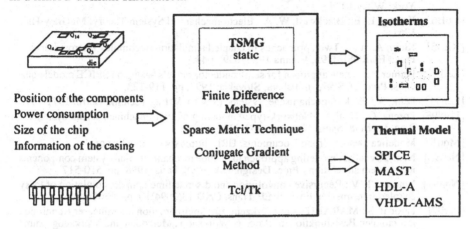

Figure 13. A model generator based on FEM/FDM models

5 CONCLUSIONS

A mathematical approach has been developed which leads to a "canonical" form of the describing equations which are applicable to modeling of microsystems. Continuous and discrete equations may be combined together. Therefore mixed-mode modeling is considered also. The canonical description may be formulated in modern hardware description languages and may be simulated by circuit and systems simulators like Saber, ELDO, and Spectre. It proved valuable in building-up libraries of microsystem components and in developing model generators for special classes of microsystem components.

References

[BoD87] Box, G.E.P.; Draper, N.R.: Empirical Model Building and Response Surfaces. Wiley, 1987.

[Cel91] Cellier, F. E.: Continous System Modeling. Berlin: Springer-Verlag, 1991.

[CHS96] Clauss, C.; Haase, J.; Schwarz, P.: An approach to analogue behavioural modelling. Proc. VHDL User Forum Europe, Dresden, 1996, pp. 85-96.

[Ecc96] Eccardt, P.C. et al.: Coupled finite element and network simulation for microsystem components. Proc. MICRO SYSTEM Technologies (MST'96), VDI-Verlag, Potsdam 1996, 145-150.

[GeD97] Gerlach, G.; Dötzel, W.: Grundlagen der Mikrosystemtechnik. Hanser-Verlag, München 1997.

30

[Ise92]   Isermann, R.: Identifikation dynamischer Systeme 2. Berlin - Heidelberg: Springer-Verlag, 1992.

[JaC92]   Jackson, M. F.; Chua, L. O.: Device modeling by radial basis functions. IEEE Trans.CAS-I 39(1992)1, pp. 19-27.

[KaR74]   Karnopp, D. C.; Rosenberg, R. C.: System Dynamics: A Unified Approach. New York: Wiley 1974.

[KoB61]   Koenig, H. E.; Blackwell, W. A.: Electromechanical System Theory. McGraw-Hill, 1961.

[Kle95]   Klein, A. et al.: Two approaches to coupled simulation of complex microsystems. Proc. EUROSIM '95, Vienna 1995, 639 - 644.

[Lei97]   Leitner, T.: A new approach for semiconductor models basing on SPICE model equations. Proc. ECS'97, Bratislava, Slovakia, 1997, pp. 119-123.

[Len71]   Lenk, A.: Elektromechanische Systeme (3 vol.). Verlag Technik, Berlin 1971-1973.

[LoN98]   Lorenz, G.; Neul, R.: Network-typed modeling of micromachined sensor systems. Proc. MSM98, Santa Clara, pp. 233-238.

[Mod]   Modelica Design Group Documents. URL: http://www.modelica.org/

[Neu98]   Neul, R. et al.: A modeling approach to include mechanical microsystem components into system simulation. Proc. Design, DATE'98, Paris, 1998, pp. 510-517.

[Ngu94]   Nguyen, T. V.: Recursive convolution and discrete time domain simulation of lossy coupled transmission lines. IEEE Trans. CAD 13(1994)10, pp. 1301-1305.

[Par97]   Parodat, S.: MARABU - Ein Werkzeug zur Approximation nichtlinearer Kennlinien mit radialen Basisfunktionen. Proc. 6. Workshop „Methoden und Werkzeuge zum Entwurf von Mikrosystemen", Paderborn, Dezember 1997, pp. 49-58.

[Rei85]   Reibiger, A.: On the terminal behaviour of networks. Proc. ECCTD '85, Prague, September 1985, pp. 224-227.

[ReS76]   Reinschke, K.; Schwarz, P.: Verfahren zur rechnergestützten Analyse linearer Netzwerke. Akademie-Verlag, Berlin 1976.

[SAW97] Senturia, S.; Aluru, N. R.; White, J.: Simulating the behavior of MEMS devices: computational methods and needs. IEEE Trans. Computational Science & Engineering, January 1997, 30-54.

[Sch84]   Schwarz, H. R.: Methode der finiten Elemente. Stuttgart: Teubner, 1984.

[Sch93]   Schwab, A. J.: Begriffswelt der Feldtheorie. Berlin: Springer-Verlag, 1993.

[SzR98]   Szekely, V.; Rencz, M.: Fast field solver for thermal and electrostatic analysis. Proc. DATE'98, Paris 1998, 518-523.

[Tho90]   Thoma, J. U.: Simulation by Bondgraphs. Berlin: Springer-Verlag, 1990.

[VHDL]   IEEE DASC 1076.1 WG Documents. URL: http://www.vhdl.org/analog/

[VoH95]   Voll, I.; Haase, J.: Rekursives Faltungsmodell für ein allgemeines Netzwerksimulationsprogramm. 40. Intern. Wiss. Kolloquium, Ilmenau, 1995, vol. 3, pp. 269-274.

[VoW97]   Voigt, P.; Wachutka, G.: Electro-fluidic microsystem modeling based on Kirchhoffian network theory. Proc. Transducer '97

[Wac95]   Wachutka, G.: Tailored modeling: a way to the 'virtual microtransducer fab' ? Sensor and Actuators A 46-47 (1995), pp. 603-612.

[WCS97]   Wünsche, S.; Clauß, C.; Schwarz, P.; Winkler, F.: Microsystem design using simulator coupling. Proc. ED&TC '97, Paris, 1997, pp. 113-118.

[Wün98]   Wünsche, S.: Ein Beitrag zur Einbeziehung thermisch-elektrischer Wechselwirkungen in den Entwurfsprozeß integrierter Schaltungen. Dissertation TU Chemnitz, 1998.

[ZiT89]   Zienkiewicz, O. C.; Taylor, R. L.: The Finite Element Method (2 vol.). McGraw-Hill, New York 1989 and 1991

VHDL-AMS, a unified language to describe Multi-Domain, Mixed-Signal designs. Mechatronic applications

V. Aubert and S. Garcia Sabiro

ANACAD/Mentor Graphics Corp., 11 A ch. de la Dhuy, 38240 Meylan (France)

Key words: VHDL-AMS, IEEE 1076.1, Multi-Domain, Mixed-Signal, Mechatronic, Simulation, EDA, Hardware Description Language, HDL

Abstract: In this paper, we present some methodology for the use of VHDL-AMS on mechatronic applications. The methodology is highlighting that VHDL-AMS is a unified Mixed-Signal language including VHDL digital features to describe pure analog designs. Without reducing the scope of the language, the subset proposed with this methodology uses at maximum the potential of classical mechanical, hydraulic and electrical simulators. Descriptions and simulation results are also presented to show the usability of such language for mechatronic applications.

1.   INTRODUCTION

VHDL-AMS 1076.1 *[1]* has just been accepted by the IEEE process as an IEEE standard. But what is it?

In the first part of this paper, we present you a quick overview of the language and in particular how it might describe mechatronic applications. The example of the TOOLSYS European project *[3]* will be taken in order to show how a subset of the analog part of this language has been chosen to allow descriptions to be simulated either on a mechanical [4], on an fluidic [5] or on an electrical simulator. But VHDL-AMS, which is the extension of the existing VHDL-93 language, has been designed to not duplicate existing VHDL functions. Thus, we will also see through some examples that the

31

J. Mermet (ed.), Electronic Chips & Systems Design Languages, 31–42.
© 2001 *Kluwer Academic Publishers.*

modeling of analog systems may require the description of both continuous (analog) and event driven (digital) behaviour.

In the second part of this paper, you will see the simulation results of some mechatronic applications using the VHDL-AMS simulator Mentor Graphics Corporation is developing.

In this paper, the term VHDL-AMS will be used for 1076.1-99, that includes 1076-93 (VHDL-93).

2. METHODOLOGY OF USE

VHDL-AMS allows users to write Differential and Algebraic Equations (also called DAEs) in several ways *[1]*. The basic form is an equation expressing the equivalence of the two expressions. This is called a *simple simultaneous statement*. Other statements allow conditional switch between sets of equations. They use an *if* or a *case control structure* to switch the sets of equations. VHDL-AMS also provides the possibility to write sequential code to compute quantities instead of using the equation notation. These statements are called *simultaneous procedural statements*.

2.1 Way to write equations

The four forms of equations seen above allow the user to describe a design having any kinds of constraints. But for mechatronic applications, it is also important to have real time simulation. This necessitates subsetting the language; so, it is a better fit to existing real-time mechanical and hydraulic simulators. In the TOOLSYS project, it has been decided to write *simple simultaneous statement* using one of the three following forms:

```
Q == f();
Q'Dot == g();
0.0 == h();
```

where there are neither Q'Dot (derivative of Q referring to time) nor Q'Integ (integration of Q referring to time) implicit quantities in the right part of the equations.

In addition, structured *simultaneous statements* have to be balanced. It means that for each clause of the *simultaneous if* or *case statements*, there must be the same number of equations and, in the case of a *simultaneous if statement*, an *else* clause is mandatory, and in the case of *simultaneous case statement*, all the possible values switching the *case* must have a clause.

if <condition> use	case <Bit_value> use
Q1 == ...;	when '0' =>
Q2 == ...;	Q1 == ...;

else
    Q1'Dot == ...;
    0.0 == ...;
end use;
	Q2 == ...;
	when '1' =>
	    Q1'Dot == ...;
	    0.0 == ...;
	end use;

This way to describe sets of equations will not significantly reduce the capabilities of the VHDL-AMS language, because it is almost always possible to describe an analog system with it, by adding new declarations and new equations for intermediate derivative quantities.

In conclusion, using the subset shown above to describe DAEs helps designers involved in the TOOLSYS project to switch easily from one simulator to another, using basic optimized features which are provided by each simulator (mechanical simulator, hydraulic simulator or electrical simulator). In addition, a tool only optimizing these kinds of equations may be developed for real time simulation.

2.2   Mixing analog and digital feature for purely analog designs

There are two aspects to the "mixed" in mixed signal. One aspect is the mixing of modeling abstractions with different mathematical foundations. So called "digital" simulators use a discrete event queuing model for simulating salient features of electronics with high computational efficiency. The assumption is that the simplified model of switches, delays, and memory elements is sufficient to predict important features to the finished silicon. Analog simulators use a different mathematical foundation – Differential / Algebraic Equations, also called DAEs.

Most interesting systems are not modelled as a single set of DAEs. Different sets of equations are used in different volumes of the state space and in different intervals of time. A set of equations is initialized and the trajectory is calculated until its coordinates reach the boundary of the volume or the time interval has elapsed. Then, a new set of equations is selected based on the current coordinate and the time. If the new equations introduce a discontinuity in any unknown or its derivative, the new equations are initialized before continuing. This sequence of events is repeated as simulation progresses.

In this paragraph, we will refer to the second aspect of the "mixed" definition.

Describing such systems using VHDL-AMS requires using the *digital part* of the language (digital signals). This is illustrated in the following example.

Let's suppose that we want to simply model the fluid flow between two chambers via an orifice. Its pressure and its flow rate characterize the fluid.

The pressure of the fluid is defined inside each chamber. The flow rate of the fluid through the orifice is computed regarding to the difference of the pressure between the two chambers.

At the beginning, the first chamber contains one fluid and the second one is empty. When one part of the fluid goes from one chamber to another, we can differentiate two successive states within the orifice:

- Laminar
- Turbulent

For each state, the set of equations is different and a VHDL digital process follows the evolution of the flow to determine what is the current state. Here is a description of such an orifice.

```
use disciplines.
    fluidic_system.all;
entity orifice is
  generic(d : Real := 5.0e-2;
    alpha  : Real := 0.7;
    recrit : Real := 1.0));
  port(terminal t1, t2 :
               Fluidic);
end entity orifice;

library IEEE;
use IEEE.math_real.all;
architecture type1 of
                 orifice is

  constant rho : Real
       := 850.0;
  constant area : Real
       := Math_Pi*d**2/4.0;
  constant nu : Real
       := 50.0e-6;
  constant pcrit : Real
       := rho*0.5 *
  (Recrit*nu/(alpha*d))**2;
  quantity deltap across
           Q through t1 to t2;
  -- pressure and flow rate
  -- between t1 and t2
  quantity Re : Real;

  function compute_q
(deltap : Real)
  return Real is
    variable q : Real;
begin
  q := alpha*area*
      sqrt(2.0*abs
deltap/rho);
    if deltap > 0.0 then

type status is
(laminar, turbulent);
-- type defining the diffe-
-- rent states of the flow

  signal current_status :
   status;
-- signal describing the
-- current state of the
-- flow

begin

  flow_rate :
  if current_status =
laminar use
    Q == -Recrit*area*nu*
         deltap/(pcrit*d);
  else
    Q == compute_q(deltap);
  end use;

  Re == abs Q * d/(area*nu);

status_synchro :
  break on current_status;
-- discontinuity handling

flow_status : process
  begin
    if abs deltap >= pcrit
    then
      current_status <=
      turbulent;
    else
      current_status <=
laminar;
    end if;
    wait on
```

`q := -q;`
end if;
`return q;`
end function `compute_q;`
	`deltap'Above(pcrit),`
	`deltap'Above(-pcrit);`
	end process;
	
	end architecture `type1;`

This example highlights different points:

- The type *STATUS* and the signal *CURRENT_STATUS* of type STATUS exhibit the state of the flow between the two chambers.
- A process called *FLOW_STATUS* has been written to observe when the difference of pressure between the two chambers crosses a critical value called *PCRIT* (in any direction). When this occurs, the value of *CURRENT_STATUS* is updated.
- Another process, called *STATUS_SYNCHRO*, resets the analog solver by signaling a discontinuity each time *CURRENT_STATUS* changes.
- The result will be, for the analog solver, to use the other set of equations in the *FLOW_RATE simultaneous if statement* each time *CURRENT_STATUS* changes.

In conclusion, it is very important to understand that VHDL'93 has been extended to VHDL-AMS by re-using as much as possible what already existed. In particular, the VHDL way to manage the evolution of the different states of a state-machine have been re-used. This way of designing using VHDL has to be re-used with VHDL-AMS to optimize descriptions.

3. EXAMPLES AND SIMULATION RESULTS

The two following descriptions illustrate multi-domain designs; the valve-solenoid model uses design entity boundary elements of electromagnetic and mechanical domains and the wheel cylinder model includes fluidic and mechanical domains. The equations describing these design entities mix energy domains.

The two examples use the methodology presented above and show the power of VHDL-AMS to describe, with one language, not only multi Domain designs, but also multiple Levels (System level of a Wheel Cylinder, Macromodel level of the electromagnetic part of a solenoid valve controlled injector) and mixed Signal designs (the different states of a brake system are managed using a digital state machine and are mixed with the DAEs defining the wheel cylinder).

36

3.1 Multi-Domain Mechanical and Electromagnetic systems: a solenoid valve-controlled injector

This example represents one injector of a common rail injector system. The injector controls the fuel flowing into the combustion chamber of the engine. It is illustrated in figure 1. A solenoid (or electromagnet) valve controls this injector.

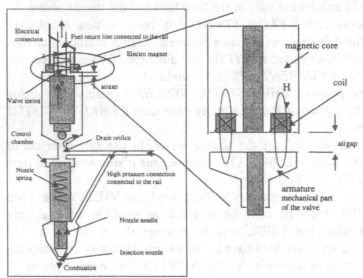

The voltage across the electromagnet induces a magnetic flux. When this flux is high enough, the solenoid pulls open the drain orifice. When the drain orifice opens, the pressure in

Figure 1

the valve control chamber drops so that the closing force on the injection nozzle is reduced and the nozzle needle opens the injection nozzle. Then, the high pressure pump delivers the fuel and the fuel flows into the combustion chamber.

Now, we describe the solenoid valve. It is composed of four parts: a coil (electromagnet), an airgap, a magnetic core and the armature of the mechanical part of the valve (see the schematic above). The schematic below (figure 2) presents the way the electromagnet has been described using a VHDL-AMS model.

In the VHDL-AMS model, the electromagnet has three conservative connections:

- one *electrical* connection (*E1*) which describes the electrical input of the electromagnet. This connection is represented by one terminal, where the *ACROSS* aspect represents the voltage (*V1*) and the *THROUGH* aspect represents the current (*I1*).

- one electromagnetic connection (*M1*) which describes the magnetomotive force and the magnetic flux provided to the magnetic core (another VHDL-AMS model). This connection is represented by one terminal, where the *ACROSS* aspect represents the magnetomotive force (*V2*) and the *THROUGH* aspect represents the magnetic flux (*I2*).
- one mechanical connection which describes the position of the armature according to the position of the magnetic core and the force acting on the armature of the solenoid. This connection is represented by one terminal, where the *ACROSS* aspect represents the distance between the armature and the magnetic core (*OFFSET*) and the *THROUGH* aspect represents the force (*FORCE*) acting on the armature.

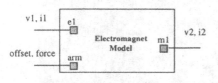

Figure 2

Thus the VHDL-AMS description is:

`entity sole2 is`
`  generic(res : real := 4.0;`
`    turn : real := 20.0;`
`    s1   : real := 6.0e-5;`
`    s2   : real := 6.0e-5;`
`    maxg : real := 1.0e-4;`
`    ming : real := 4.0e-5);`
`  port(terminal e1 :`
`Electrical;`
`    terminal m1 : Magnetic;`
`    terminal arm :`
`Kinematic);`
`end entity sole2;`
	`architecture arch1 of sole2 is`
	
	`quantity v1 across`
	`        i1 through e1;`
	`quantity mmf across`
	`        flux through m1;`
	`quantity offset across`
	`        force through arm;`
	`quantity dphi, mmfair, gap :`
	`  Real;`
	
	`constant a_area : Real`
	`  := 1.0/s1+1.0/s2;`
	
	`begin`
	
	`gap == maxg-offset;`
	`mmfair == -flux/u0*a_area*gap;`
	`mmf == i1*turn - mmfair;`
	`dphi == (v1 - i1*res)/turn;`
	`flux'Dot == -dphi;`
	`force ==`
	`        0.5*flux**2 / u0*a_area;`
	
	`end architecture arch1;`

Simulations of the description above were performed using a VHDL-AMS simulator developed by Mentor Graphics Corporation.

The input voltage (*V1*) of the electromagnet controls the resulting displacement (*OFFSET*) of the mechanical part of the valve and the resulting electromagnetic force (*FORCE*) applied on the mechanical part of the valve (figure 3).

38

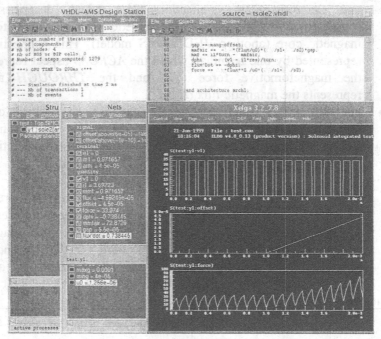

Figure 3

The peaks of the magnetic force correspond to the maxima of the voltage source. When the offset increases, the mechanical part of the valve is moving from bottom to top and the airgap decreases; In this case, the magnetic force increases.

3.2 Mechanical system: a wheel cylinder, part of a full ABS brake engine

This second example represents a part of an ABS brake system. What will be modeled using VHDL-AMS is the wheel cylinder [2], as described in the following schematic (figure 4). When the driver brakes, the pressure in the pipe pushes a piston which generates a torque on the wheel in order to brake it.

The wheel cylinder receives as one input the hydraulic pressure provided by the ABS system. It reacts by pushing the brake pads onto the brake disk. This action is done in two steps:

- No brake action as long as the brak pads are not in contact with the disk, but they may advance to this contact
- Brake action when contact made and pressure continues to be provided by the ABS system.
- To described it using VHDL-AMS, we have defined three brake states:

✓ *inactive*: no brake action has been provided
✓ *advance*: brake action has been started but the break pads are not in contact with the brake disk (no action on the speed of the wheel, the torque provided by the wheel cylinder is equal to 0.0)
✓ *contact*: brake action continues and the brake pads are in contact with the brake disk. The wheel locking may occur if the torque is large enough.

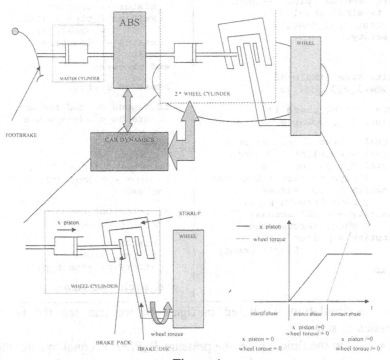

Figure 4

In the VHDL-AMS model, the wheel cylinder has two conservative connections:

- one *fluidic* connection which describes the pressure and the flow rate coming from the ABS system. This connection is represented by a terminal, where the *ACROSS* aspect represents the pressure and the *THROUGH* aspect represents the flow rate out of the wheel cylinder.

- one *mechanical* connection which describes the angular velocity of the wheel and the torque provided by the wheel cylinder. This connection is represented by a terminal, where the *ACROSS* aspect represents the angular velocity and the *THROUGH* aspect represents the torque from the wheel cylinder acting on the wheel.

Thus, a VHDL-AMS description may be:

```
entity wheel_cylinder is
  generic(
    P_contact : Real := 5.0e4;
    P_inactive : Real := 1.0e3;
    S_piston  : Real := 1.0e-3;
    friction_coeff
            : Real := 1.0e-2;
    piston_stroke
            : Real := 1.0e-3);
  port(terminal pipe : Fluidic;
    terminal wheel
    Rotational_omega);
end entity;


architecture simple of
    wheel_cylinder is

  type Braking_phase is
   (inactive, advance, contact);

  signal status : Braking_phase;
  signal w_wheel_ab : Boolean;
  signal P_pipe_inactive,
      P_pipe_contact : Boolean;
  quantity P_pipe across
      Q_pipe through pipe;
  quantity w_wheel across
      C_wheel through wheel;
  quantity x_piston :
                Displacement;

begin
```

```
  w_wheel_ab <=
      w_wheel'Above(0.01);
  P_pipe_contact <=
      P_pipe'Above(P_contact);

  process
  begin
    wait on
      P_pipe'Above(P_inactive);
    status <= advance;
    wait on P_pipe_contact;
    status <= contact;
    wait on P_pipe_contact,
            w_wheel_ab;
    status <= inactive;
    wait;
  end process;

  break on status;

  if status = inactive or
     status = advance use
     C_wheel == 0.0;
  else
     C_wheel ==
  friction_coeff *
  (P_pipe - P_contact);
  end use;

  x_piston'Dot ==
  Q_pipe / S_piston;

  x_piston ==
    piston_stroke *
    atan(P_pipe*math_pi*1.0e-5);

end architecture;
```

In the simulation illustrated in figure 5, we can see the following succession of states:

- As in the fluidic pipe, the pressure *P_PIPE* is equal to zero, there is no brake action and the state of the brake corresponds to *inactif*.
- As the pressure *P_PIPE* is superior to 1.0e3 bar but inferior to 5.0e4 bar, the state of the brake is *avance*. The position of the piston *X_PISTON* increases.
- When the pressure *P_PIPE* is superior to 5.0e4 bar, the break pads are in contact to the brake disc and the brake state is *contact*. The torque *C_WHEEL* increases until *P_PIPE* increases and becomes constant when *P_PIPE* is constant.
- As the angular velocity of the wheel *W_WHEEL* becomes equal to zero the state of the brake returns to the *inactive* and the torque becomes equal to zero.

The output displays are:

- *P_PIPE* and *Q_PIPE* which are the hydraulic pressure and the flow rate in the pipe connected to the ABS

- *W_WHEEL* and *C_WHEEL* which respectively correspond to the angular velocity of the wheel and to the torque getting from the wheel cylinder and acting on the wheel
- *X_PISTON*, an intermediate state which represents the displacement of the piston
- *STATUS* which is a digital signal representing the state of the brake
- The presented description is interesting because it allows to show how the different states of the brake are handled but the release of the brake pedal before engaging the break pad can also be taken into account.

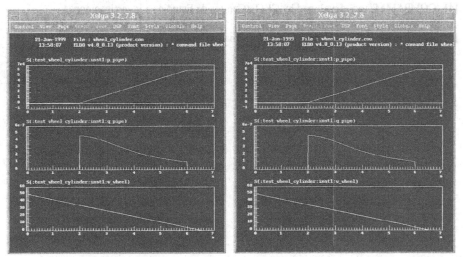

Figure 5

4. CONCLUSION

This paper presents the new language called VHDL-AMS, illustrating it with mechatronic examples. VHDL-AMS has been extended from VHDL and has just been accepted as an IEEE standard and can handle Mixed-Signal, Multi-Domain and Multi-Level descriptions. Simulations results have also been presented using the VHDL-AMS simulator Mentor Graphics Corporation is building.

We can also see through these examples that it is possible to build mechatronic component libraries to be reused for designing automotive systems.

In addition, an important point consists of the simulation of the entire ABS system or the entire Common Rail system. Through these examples, it has been shown that fluidic, mechanical and electromagnetic systems can be simulated together. In conjunction with the electrical level (SPICE description in Mentor Graphics VHDL-AMS simulator) and the digital VHDL level, it should be possible to simulate the entire systems.

1. REFERENCES

[1] "IEEE Standard VHDL Analog and Mixed-Signal Extensions, IEEE Std. 1076.1-1999

[2] "Automotive Handbook, 4th Edition", published by Robert Bosch GmbH, distributed worldwide by SAE, 1998

[3] "TOOLSYS project, BE 96-3249", BRITE European project, 1996

[4] "COMPAMM" tool developed by CEIT (Centros de Estudios e Investigacionnes Technicas) in San Sebastian Spain

[5] "AMESim" tool developed by Imagine in Roanne France

Efficient Modeling of Analog and Mixed A/D Systems via Piece-wise Linear Technique

Jerzy Dąbrowski, Andrzej Pułka
Institute of Electronics, Silesian University of Technology, Gliwice, Poland

Key words: Piece-wise linear modeling, Functional-level modeling, VHDL models

Abstract: *In this paper an application of the piece-wise linear (PWL) modeling
technique is presented. It is oriented towards functional-level modeling of
analog and mixed-signal A/D systems. The principles of the actual PWL
approach are briefly described. The models are based mainly on first order
differential equations, which are solved by an explicit PWL algorithm. All the
analog signals propagating through a network are PWL signals obtained with
the built-in approximator. Examples of some models are discussed in detail,
including a support by the waveform relaxation technique. Implementation of
the PWL models in a discrete VHDL environment is emphasized. Some
simulation examples of A/D networks, obtained by means of the V-System
simulator are also included.*

1.   INTRODUCTION

Due to the current proliferation of ICs the need of more and more
powerful design tools is becoming apparent. Great effort has already been
made to cover this need in digital domain. In the analog domain, however,
circuits are usually designed and verified at the device level, despite the fact
that device verification level is very CPU intensive. In particular, the analog
or mixed A/D simulation of large systems performed at the, so called,
SPICE-level suffers from an excessive amount of computer. Hence,
following the top-down design strategy, the analog modeling and simulation
adequate also for higher levels of abstraction are actually important
objectives. On the other hand, high-level, rough models are usually no more
sufficient. Consequently, a new generation of models (e.g. functional level)

43

J. Mermet (ed.), Electronic Chips & Systems Design Languages, 43–54.
© 2001 *Kluwer Academic Publishers.*

tends to comprise more specifications, in order to provide a designer with more detailed verification results prior to step down to lower design levels. Clearly, timing specifications are of particular interest in most cases. Since the respective timing models are usually based on differential equations, their typical implementation requires very CPU-time consuming procedures.

Recently, a new piece-wise linear (PWL) approach has been proposed to cope with this problem [6]. It can be applied to analog or mixed analog/digital networks represented mainly at the functional level. The PWL approach features: the PWL signals, inertial building blocks as basic modeling units and the explicit simulation algorithm to solve for network equations. For digital units, however, a behavioral description is preferred. As a consequence, the PWL-to-logic and logic-to-PWL converters are required when modeling AD systems.

In this paper we present the PWL modeling and simulation technique of analog and mixed AD systems. Mainly, the functional level is considered. In Section 2 we derive briefly the actual PWL approximation algorithm with respect to the basic building blocks used for synthesis. In Section 3 we give insight into the PWL modeling of analog units, like amplifiers with nonlinear effects or active filters. The problems arising for models with, so called, strong feedback loops are emphasized, and the convergence of the PWL algorithm is discussed for this case. Section 4 addresses the implementation of PWL models in VHDL environment. Two approaches to PWL modeling are considered. Besides, we include also the simulation results for two mixed A/D networks, obtained by means of V-System.

2.   BASICS OF PWL MODELING TECHNIQUE

In the PWL modeling signals can be represented by means of subsequent points (t_k, V_k), (t_{k+1}, V_{k+1}), ..., where t_k is the k-th time instant and V_k stands for the respective voltage amplitude. Defining the local rate as $s_k = (V_{k+1} - V_k)/(t_{k+1} - t_k)$, the k-th PWL segment may be expressed as: $V(t) = V_k + s_k(t - t_k)$, and when applied to the input of the integrator or inertial block it is converted to a smooth, nonlinear curve at the output. In fact, analog networks consist of subcircuits that often exhibit inertial properties. Therefore, any analog unit can usually consist of a few inertial building blocks to mimic the timing behavior, the basic nonlinearities (e.g. saturation) and the output loading effects. The constitutive relation of the basic building block takes the form of:

$$T\frac{dx}{dt} + x = f(x_{inp}) \tag{1}$$

where $f(.)$ denotes its DC characteristic and T its time constant (Fig.1).

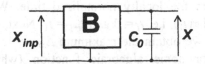

Figure 1. Basic building block as two-port

When capacitive loading effects at the output x must be accounted for, the time constant comprises the capacitance C_0. Clearly, eqn. (1) can be generalized into a multiple input case, e.g. for a multiplier or adder.

Using the mentioned above signal V as x_{inp}, and assuming the PWL approximation for $f(.)$, a PWL signal $u(t) = f[V(t)]$ is obtained. For $t_k = 0$ we have: $u(t) = u_0 + r_0 t$, $t \in [0, t_{max}]$ $(t_{max} \leq t_{k+1})$. Hence, solving for (1) an explicit formula for x, consisting of the transient and steady state component follows:

$$x(t) = (x_0 - u_0 + r_0 T)\exp(-\frac{t}{T}) + r_0(t - T) + u_0 \qquad (2)$$

The main objective here is to get a PWL approximation of (2) to enable further propagation of the signal x in a linearized form, so that when a system is modeled all the links between analog units take a form of PWL signals. For this purpose we first split the time interval $[0, t_{max}]$ into subintervals $[0, t_1]$, $[t_1, t_2]$, ...$[t_n, t_{max}]$. For each subinterval ...$[t_i, t_{i+1}]$ a segment of a PWL approximating signal x_{lin} is defined by its end points that are assumed to lie on the curve x. Hence, we have: $x_{lin}(t_i) = x(t_i)$ and $x_{lin}(t_{i+1}) = x(t_{i+1})$. In fact, given t_i, the time t_{i+1} has to be found (Fig.2). To control the accuracy of this approximation the Chebyshev measure has been found the most advantageous. Consequently, our objective can be formulated as an optimization task that is to maximize the distance $d = t_{i+1} - t_i$ with some constraints (assumed approximation accuracy p_{mx}) and given t_i.

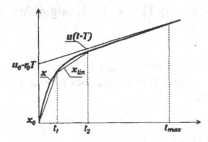

Figure 2. PWL approximation of output signal

The actual approach to solve this problem is a direct algorithm [14], which makes use of precise formulas, derived from the pure maximum condition for the Chebyshev distance function:

$$\Delta_x(\tau) = |x(\tau) - x_{lin}(\tau)| \qquad (3)$$

where $\tau = t/T$ denotes the locally normalized time. We assume that x_{lin} crosses through x_0 and $x(t_1)$, i.e. $t_i = 0$ and $t_{i+1} = t_1$. Next, from the maximum condition: $d\Delta_x/d\tau = 0$ we obtain the maximum $\Delta_{xmx} = |x_0 - u_0 + r_0 T| \phi(\tau_1)$, where $\phi(.)$ stands for some algebraic function (which asymptotically approaches 1) and $\tau_1 = t_1/T$. By letting $\Delta_{xmx} = p_{mx}$, the required relation between the approximation accuracy and the normalized segment length τ_1 is obtained (Fig.3). In practice, the reciprocal of the ϕ function is needed, i.e. $\Phi = \phi^{-1}$, so that the approximation task may be solved directly based on the relative accuracy p_{mxr}

$$\tau_1 = \Phi(p_{mx}), \qquad p_{mxr} = p_{mx} / |x_0 - u_0 + r_0 T| \qquad (4)$$

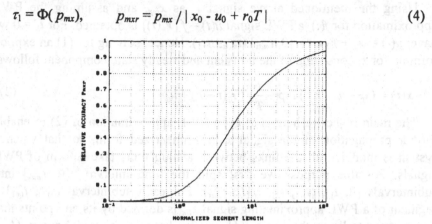

Figure 3. Relative accuracy p_{mxr} versus normalized segment length τ_1

Using (4) the subsequent approximation points can be achieved assuming that each PWL segment begins at $t_i = 0$ and ends at $t_{i+1} = \tau_1 T$. It follows that x_0 and u_0 have to be updated for each PWL segment, i.e. the next x_0 can be found from (2) by means of the substitution: $x_0 \leftarrow x(\tau_1 T)$, and the next u_0 in the same way: $u_0 \leftarrow [u_0 + r_0 \tau_1 T]$. A PWL algorithm given below may summarize this discussion:

```
repeat
    if x_0 - u_0 + r_0 T <> 0 and x_0 - u_0 + r_0 T > p_mx
        then { τ_1 ← Φ(p_mxr ); if τ_1 > τ_mx then τ_1 ← τ_mx }
        else τ_1 ← τ_mx ;
    x_k ← x(τ_1 T);
    x_0 ← x_k ; u_0 ← u_0 + r_0 τ_1 T;
    t_k ← t_{k-1} + τ_1 T;
    k ← k + 1;
    τ_mx ← τ_mx - τ_1 ;
until τ_mx = 0
```

The result of this procedure are two sequences: $\{t_k\}$ and $\{x_k\}$ that define the PWL output waveform of any inertial block for a single input segment u. The values of the Φ function are calculated by table look up and linear interpolation. The same method is used for the exponential function, when computing subsequent values of xk. The full discussion of the algorithm and its properties is given elsewhere [14].

3. PWL ANALOG MODELS

The PWL technique is particularly well suited to analog modeling at the functional level. The building blocks used for synthesis are assumed to be unidirectional, so that currents are usually not accounted for. Consequently, electrical effects that require bi-directional signal flow (e.g. in case of a transmission gate) or tight feedback loops should be avoided by incorporating them into the building blocks.

The functional-level PWL models of analog units, like amplifiers, voltage comparators, or D/A converters, proved to be computationally efficient and relatively accurate, when compared to the respective SPICE estimates [6,7]. Here, we address only the model of a noninverting amplifier as an illustration. The amplifier specifications taken into account are as follows: gain, dominant pole, saturation, output impedance and the slew rate.

For small input/output signals a fully linear model is sufficient. However, to cover a full range of input amplitudes the nonlinear function $f(.)$ and the slope limiting mechanism (SLM) must be used. The SLM does not begin to act until çDuinç> uth, where $Duin$ is an initial increment of the input signal (when starting from he quiescent point) and uth denotes the threshold voltage that puts the amplifier into saturation. Each segment of uin with amplitude exceeding this threshold is checked for the slew rate (SR) parameter. The model consists of the SLM and two cascaded inertial blocks. The first gives the gain, the dominant pole and the output clamping, whereas the other one serves as an output stage. The inverse the time constant ($T0$) of the first block is the dominant pole frequency $w0$, and the time constant of the second block comprises the output resistance $Rout$ and output capacitance $Cout$ (including a capacitive load).

In Fig.4 the functioning of the SLM for a large input is shown (uin consists of three segments). An amplifier with the following parameters is considered: $k=10$, $w0=125600$rad/s (f1 =1Mhz), $SR=0.5$ V/ms. The slope of the first input segment is limited after crossing the uth =80mV. When the output of the SLM (dotted line in Fig.4) reaches the value of 0.4V, the difference between in (solid line) and the feedback signal $uout/k$ (dashed line) is checked. Since in this case it is approximately 160mV, the amplifier input stage is still saturated (uth =80mV) and the amplifier continues to slew. To model this effect, the SLM continues to rise with no change of slope until the uin is crossed. Next, it follows the uin segments.

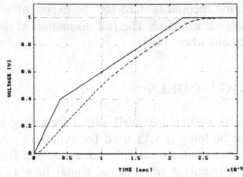

Figure 4. Functioning of slope limiting mechanism for large input (k = 10)

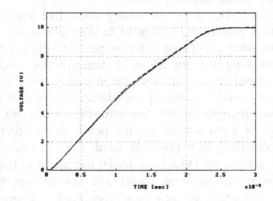

Figure 5. Amplifier time responses for k = 10

In Fig.5 the corresponding amplifier time response is given. For comparison the SPICE estimate is plotted with the dashed line. The PWL simulation of this model is up to three orders of magnitude faster than SPICE depending on the approximation accuracy p_{mx}.

Most of the analog and mixed A/D models works well, based on the presented PWL approximation algorithm. The obtained, discrete points (t_k, V_k) are defined to be the events that control the simulation process using the 'next event' technique [8]. Some networks, however, are provided with tight feedback loops, that cannot be avoided. That means, they must be modeled to represent adequately the network behavior. As a consequence, a problem of iterations arises. Fortunately, the PWL technique supported by the waveform relaxation (WR) proved to cope well with this drawback [13]. Observe that the PWL technique is well suited to the WR, since the PWL signal segments are, in fact, simple waveforms defined by their boundary points. Although WR is said to suffer from slow convergence, when tight feedbacks are present, the resulting in this case CPU times are fairly moderate. It is because the PWL algorithm is very time effective, and additionally, in some cases the number of iterations can be substantially reduced with the windowing technique [9].

Here, to give insight into this approach we use the simple PWL model of an analog filter. A second order, low pass filter composed of the subtractor, integrator and inertial block will be considered (Fig.6).

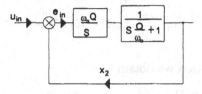

Figure 6. Model of bi-quad low-pass filter

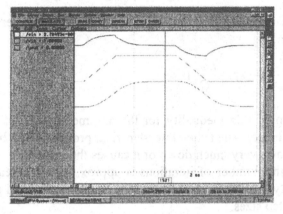

Figure 7. Waveforms of bi-quad low-pass filter simulation with V-System

Apparently, it has got a tight feedback loop, and hence, when simulated it requires iterations. Some of the simulation results for the low pass filter obtained with the VHDL simulator are given in Fig.7. The specifications for the filter, defined according to Bessel are as follows: $Q=0.557$, $f_n=1.274$. For the cutoff frequency $f_c=1000$Hz obtains $\omega_0=2\pi f_n f_c = 8004$rd/s. The waveform relaxation algorithm is incorporated into the model. A dynamical windowing technique is used, so that each time window matches the actual PWL segment size.

For very small values of the expression: $x_0 - u_0 + r_0 T$ (for the inertial block) or r_0/T (for the integrator), relatively large time steps are produced, that are likely to make the relaxation process unstable. To discuss this problem we describe the filter structure by the following equations:

$$\frac{1}{\omega_0 Q}\frac{dx_1}{dt}=u_{in}-L(x_2), \quad \frac{Q}{\omega_0}\frac{dx_2}{dt}+x_2=L(x_1) \tag{5}$$

where x_1 and x_2 are respectively the output of the integrator and the inertial block, and $L(.)$ denotes the PWL operator. As mentioned earlier, the length of a segment propagating through the loop stabilizes after a few iterations at some value t_1, so that $L(x_1) = x_1(0) + [x_1(t_1) - x_1(0)]t/t_1$. A similar relation holds for $L(x_2)$. Consequently, for the j-th iteration we have:

$$x_1^{j+1}(t_1) = x_1(0) + \omega_0 Q \int_0^{t_1} (u_{in} - x_2(0) - \frac{x_2(t_1) - x_2(0)}{t_1} t) dt \quad (10) \tag{6}$$

$$x_2^{j+1}(t_1) = x_1(0) + \frac{x_1^{j+1}(t_1) - x_1(0)}{t_1}(t_1 - \frac{Q}{\omega_0}) + [x_2(0) - x_1(0) + \frac{x_1^{j+1}(t_1) - x_1(0)}{t_1} \frac{Q}{\omega_0}]\exp(-\frac{t_1\omega_0}{Q})$$

After simple manipulation we obtain:

$$x_2^{j+1}(t_1) - x_2^j(t_1) = -[\frac{\omega_0 Q}{2} t_1 - \frac{Q^2}{2}(1 - \exp(-\frac{t_1\omega_0}{Q}))][x_2^j(t_1) - x_2^{j-1}(t_1)] \tag{7}$$

Now, by virtue of the contraction mapping theorem the iteration process converges when:

$$\frac{\omega_0 Q}{2} t_1 - \frac{Q^2}{2}(1 - \exp(-\frac{t_1\omega_0}{Q})) < 1 \tag{8}$$

The solution of this inequality for the assumed values of ω_0 and Q is $t_1 < 0.505$ ms. Hence, when the time step t_1 approaches this boundary, the convergence slows very much down or it causes the instability. For example, for $t_1 = 0.49$ms the number of iterations is approx. 100, whereas for shorter steps it varies typically between 5 and 15. Clearly, the model must prevent such unstable step sizes.

A more comprehensive discussion pertaining to the PWL waveform relaxation and its stability may be found in [13].

4. IMPLEMENTATION OF PWL MODELS IN VHDL

There are two possible approaches to the PWL modeling of analog and mixed-signal networks in the VHDL environment. The first technique, the simpler one [12], allows to model small circuits with no tight feedback loops, but it accepts feedbacks that are cut by clocking signals. The other approach is required when the circuit being modeled is relatively complex or contains strong feedback.

For simple networks the architecture of the model can be reduced to one process without sensitivity list. This methodology requires a mechanism of the **wait** statement, which freeze the simulation till the next PWL segment. Besides, the PWL step- and amplitude calculations must be performed (Example 2). As opposed to this, the sensitivity lists are necessary for the complex models and especially when iterations have to be performed.

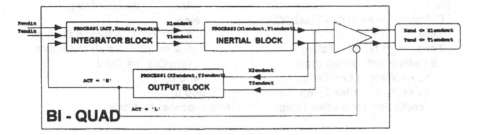

Figure 8. Structure of VHDL low-pass filter model

To address this approach, first recall the second order low-pass filter described in Section 3. As it constitutes a tight feedback loop, the iterative technique has to be used. Figure 8 presents the relevant VHDL blocks (processes) and the communication between them. Process1 - the integrator module, and process2 - the inertial module are responsible for generation of proper PWL segments. The outputs of process1 are connected to the inputs of process2. The nature of both pairs of signals (X1endout, T1endout) and (X2endout, T2endout) is the same, they represent the amplitude and time of the end of the PWL segment. The iterative effect has been achieved thanks to the third block (output block) modeled with process3. This block controls the accuracy of the iterations by comparing the difference between the subsequent PWL segments of the inertial block. When the difference exceeds the assumed accuracy, the output block activates process1 and forces the next iteration (signal ACT = 'H'). The final output PWL segment of the filter model (Yout, Tout) is also formed by this block, when the end of iterations is detected. This relatively simple solution allows performing iterations without any additional mechanisms. The natural iterations (delta cycles) built into the VHDL simulator are employed, and when convergence is assured, usually several steps have to be done for a given PWL segment.

A fragment of the VHDL code of this model is presented below and an example of the relevant PWL waveforms obtained with the V-System (ModelTech) is given in Fig.7.

-- Example 1: fragment of structural model of the bi-quad low-pass filter

```
(...)entity BiQuad is
port(
    Xstart, Xend:        in    real;
    Tend:                in    time;
    Ein, EndOut:         inout real;
    TimeStep, TendOut:   inout time);
end BiQuad;
-- description of BiQuad using
-- component instantiation statements
(...)
architecture structural of BiQuad is
```

```
-- components declarations; here:
-- integrator, inertial block and
-- output buffer
-- ports and signals instantiation(...)
end structural; (...)
architecture only of out_bufer is
begin
process(PWLend,Y2,Tstep)
    -- variable declarations
    begin
    if Initial then   -- Initialization
```

```
else                                        Ack <= 'L' AFTER Tstep;
  Difference := abs((Y2 - Ylast)/Y2);       else
  Ylast    := Y2;                             Y1 <= Ylast ; -- forcing the next
  Tend     := NOW + Tstep;                    Ack <= 'H';    --step of iteration
  if (delta > Difference) then                TendOut <= Tend;
    Y3 <= Ylast  after Tstep;                 end if;
    Y1 <= Ylast  after Tstep;               end if;
    TendOut<= Tend after Tstep;          end process; end only;
```

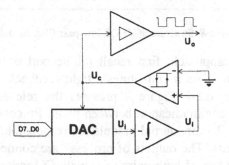

Figure 9. Structure of digit to frequency converter

The next example concerns a model of the first type - a functional-level model of the digit to frequency (D/f) converter (Fig.9). It does not contain a tight feedback, so no iterations are required for it. In fact, the signal loop is broken because of switching at the comparator output between constant +Ur and –Ur, used as a reference for DAC. The implementation of the blocks constituting this model is provided with simple processes with empty sensitivity lists. As mentioned earlier, it is at the expense of extra conditional statements incorporated into those processes (Example 2).

-- Example 2: checking of input signals transitions for integrator

```
if (Xend'event or Tend'event) then         if alpha = 0.0 then
  if not(Start) then                         T1 := DeltaTime;
    Xactual := Xlin;                        else
    -- other initialization statements       T1:= 2.0*SQRT(Pmax/abs(alpha));
  end if;                                   end if;
  T0 := NOW;                                Tstep := real_to_time(T1);
  Delta := time_to_real(Tend-T0);          Start := false;
  alpha := (Xend-Xactual)/(2.0*Delta);   end if;
```

The feedback loop in this network consists of the multiplying DAC, the inverting integrator and the voltage comparator with hysteresis. A behavioral macromodel is used for the comparator (similar to a logic gate model). On the other hand, the DAC model follows the fundamental formula $U_{out} = \Sigma 2^i a_i \Delta U$, where i= 0..7, ΔU is the resolution, and a_i parameters represent the values of input bits. Besides, for the output stage of DAC an inertial block is used.

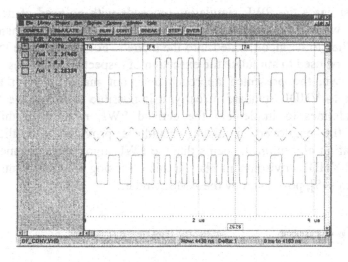

Figure 10. PWL waveforms for D/f converter from Fig.9

For constant input the output buffer delivers a square wave of a stable frequency proportional to the input digital word. Any change in the input word propagates through the feedback loop. Figure 10 shows some waveforms of the D/f converter obtained with the V-System simulator. Other examples of mixed A/D models can be found [12].

5.   SUMMARY

A modeling technique for the simulation of analog functional units in analog and mixed A/D systems has been presented. The component units are assumed to be unidirectional, but capacitive loading effects, typical of MOS circuits, are allowed. The signals are represented as piece-wise linear waveforms that may approximate closely the real timing behavior. Direct method has been used to perform the PWL approximation of smooth analog responses. Timing behavior of the derived PWL analog models matches well the respective SPICE estimates with the simulation speed-up of up to three orders of magnitude. Since the analog PWL models are of discrete nature, their implementation in VHDL environment has been feasible. However, for mixed A/D systems a mixed-mode technique is preferred, i.e. digital units are modeled behaviorally, and when interfacing with analog blocks they must be provided with logic-to-PWL and PWL-to-logic converters, respectively. Explicit formulas available for analog timing have been used to define the behavioral models. The points that create PWL waveforms are referred to as simulation events.

In most cases the PWL simulation proceeds with no need of iterations. However, when tight feedback loops must be modeled, a mechanism of iterations has to be incorporated into the models. For this purpose the delta cycles can be used to stimulate the execution of respective processes.

As shown, the presented approach does not make use of the AMS extensions to VHDL standard. Clearly, it is possible to follow the VHDL-AMS guidelines to implement the derived PWL models. In this case, however, the employed simulator should be provided with the PWL approximation algorithm. It seems that the PWL algorithm implemented in the digital VHDL environment plays a role of analog solution points (ASP) mechanism [15], postulated by the AMS standard.

REFERENCES

[1] J.M.Berge, Modeling in Analog Design, Kluwer Academic Publishers, 1995.
[2] A.Mantooth, Modeling with an Analog Hardware Description Language, Kluwer Academic Publishers, 1994.
[3] J.M.Berge, Fonkoua A., Maginot S., Rouillard J. - VHDL Designer's Reference, Kluwer Academic Publishers, 1992.
[4] J.P.Mermet, - Fundamentals and Standards in Hardware Description Languages, NATO ASI Series, Kluwer Academic Publishers, 1993.
[5] G.Ruan, J.Vlach, J.Barby, A.Opal, Analog Functional Simulator for Multilevel Systems, IEEE Trans. on CAD, vol. 10, No.5, May 1991, pp.565-575.
[6] J.Dąbrowski, Functional-Level Analog Macromodeling with Piecewise Linear Signals, *Proceedings of EURO-DAC'95*, Brighton, September 18-22, 1995, pp.222-227.
[7] J.Dąbrowski, J.Konopacki, Implementation of A/D Network Macromodels in PWL Functional-Level Simulator, Bull.of Polish Academy of Sciences, Technical Sciences, vol.44, No.3 1996, pp.293-312.
[8] R.A.Saleh, A.R.Newton, Mixed-Mode Simulation, Kluwer Academic Publishers, 1990.
[9] A.E.Ruehli, Ed., Circuit Analysis, Simulation and Design, Part 1&2, Elsevier Scence Pub. 1986, 1987.
[10] J.R.Armstrong, Chip-Level Modeling, Prentice Hall, Englewood Cliffs, N.J. 1988.
[11] IEEE Standard VHDL Language Reference Manual (Integrated with VHDL-AMS changes), IEEE Std 1076.1-1997, IEEE Standards.
[12] J.Dąbrowski, A.Pułka, Discrete Approach to PWL Analog Modeling in VHDL Environment, Analog Integrated Circuits and Signals Processing, Kluwer Academic Publishers, Vol. 16, No. 2, 1998 pp.91-99.
[13] J.Dąbrowski, Waveform Relaxation Approach to PWL Simulation of Analog and Mixed A/D Networks at the Functional Level, Proc.of ECCTD'97 , Budapest 2-6, Sept.1997.
[14] J.Dąbrowski, Functional-Level Analogue Macromodelling with Piecewise Linear Signals, IEE Proc. Circuits, Devices and Systems, vol.146, No 2, Apr.1999, pp.77-82.
[15] A.Vachoux, Analog and Mixed-Signal Extensions to VHDL, Analog Integrated Circuits and Signals Processing, Kluwer Academic Publishers, Vol. 16, No. 2, 1998 pp.185-200.
[16] R.E.Harr, A.G.Stanculescu (Ed.), Aplications of VHDL to Circuit Design, (Chapter 3 and 4), Kluwer Academic Publishers, 1991.

OO-VHDL

SUAVE: OBJECT-ORIENTED AND GENERICITY EXTENSIONS TO VHDL FOR HIGH-LEVEL MODELING

Peter J. Ashenden
Dept. Computer Science
University of Adelaide, SA 5005,
Australia
petera@cs.adelaide.edu.au

Philip A. Wilsey and Dale E. Martin
Dept. ECECS, PO Box 210030
University of Cincinnati
Cincinnati, OH, 45221-0030, USA
phil.wilsey@uc.edu, dmartin@ececs.uc.edu

1. INTRODUCTION

VHDL is widely used by designers of digital systems for specification, simulation and synthesis. Increasingly, designers are using VHDL at high levels of abstraction as part of the system-level design process. At this level of abstraction, the aggregate behavior of a system is described in a style that is similar to that of software. Data is modeled in abstract form, rather than using any particular binary representation, and functionality is expressed in terms of interacting processes that perform algorithms of varying complexity. A subsequent partitioning step in the design process may determine which aspects of the modeled behavior are to be implemented as hardware subsystems, and which are to be implemented as software.

Experience in the software engineering community has lead to adoption of object-oriented design and programming techniques for managing complexity through abstract data types (ADTs) and re-use [10]. Features included in programming languages to support these techniques are abstraction and encapsulation mechanisms, inheritance, and genericity. The term "object-based" is widely used to refer to a language that included abstraction and encapsulation mechanisms [24]. The term "object-oriented" is used to refer to a language that additionally includes inheritance.

J. Mermet (ed.), Electronic Chips & Systems Design Languages, 57–70.

While VHDL can be used for modeling at the system level, it has some deficiencies that make the task more difficult than it would otherwise be. These difficulties center around language features (or lack of some features) for supporting complexity management. VHDL is currently somewhat less than object-based, as its encapsulation mechanism are weak. It is certainly not object-oriented, as it does not include any form of inheritance. While it does include a mechanism for genericity, that mechanism is severely limited, allowing only parameterization of units by constant values. We have discussed these issues in a previous paper [3].

SUAVE aims to improve support for high-level modeling in VHDL by extending the language with features for object-orientation and genericity in a way that does not disturb the existing language or its use. As well as adding specific language features, some existing features are generalized, the facilities for encapsulation are strengthened, and an inheritance mechanism is added. Private types and private parts in packages support improved encapsulation. Type derivation, record type extension, and class-wide types with dynamic dispatching support inheritance.

SUAVE also extends the genericity mechanism of VHDL by allowing types to specified as formal generics and by allowing generics to be specified in the interfaces of subprograms and packages. Use of formal type generics allows units to be reused in contexts where data of different types is to be manipulated. For example, a multiplexer can be specified with a formal type generic for the type of the input and output data. This allows the multiplexer model to be reused as a bit, bit_vector, std_logic, std_logic_vector, integer, or user-defined-type multiplexer, without modifying the original model code. Use of generics in the interfaces of subprograms and packages allows definition of container abstract data types that can be reused to contain data of different types. For example, a generic package can be defined to represent and manipulate sequences of integer, time values, or test vectors for different devices under test, again without modifying the original package code.

We have previously argued [4] that, in addition to supporting object-orientation, these extensions improve the expressiveness of VHDL and enhance reuse across the modeling spectrum from high-level to gate-level. Furthermore, the genericity extensions interact with the extensions for object-oriented data modeling to significantly improve support for high-level behavioral modeling and for developing test-benches. By choosing an incremental and evolutionary approach to extensions, SUAVE avoids major additions to the language that would complicate choice of mechanisms for expressing a design. In addition, the implementation burden is not large, and there is no performance penalty in simulation or synthesis if the mechanism are not used.

The SUAVE approach is similar to that proposed by Mills [17] and by Schumacher and Nebel [20]. It is contrasted with others that have been proposed [9, 12, 19, 22, 25], that add new, separate mechanisms for combining abstraction, encapsulation, and inheritance for object-orientation. Such mechanisms replicate aspects of the existing features of VHDL, making design choices for expressing a model more complex.

This paper outlines the SUAVE extensions for object-orientation and genericity, and illustrates their use through some examples. More complete presentation of the extensions can be found in the SUAVE report [6].) Most of the features added to VHDL are adapted from features in Ada-95 [16], and are included largely for the same reasons that they are included in Ada-95 [8]. Section 2 of this paper outlines the design principles and objectives that were followed in deciding how to extend VHDL. Subsequent sections describe the extensions in detail and illustrate them with examples. Section 3 describes the extensions to the type system of VHDL to support type derivation, extension and class-wide programming. Section 4 describes the extensions that improve the encapsulation features of VHDL. In combination, the extensions in these two sections turn VHDL into an object-oriented language. Section 5 describes the SUAVE type genericity extensions. Our conclusions are presented in Section 6.

2.   SUAVE DESIGN OBJECTIVES

A previous paper [4] reviews the issues to be addressed in extending VHDL for high-level modeling and discusses principles that should govern the design of language extensions. As a result of that analysis, a number of design objectives were formulated for SUAVE:
• to improve support for high-level behavioural modeling by improving encapsulation and information hiding capabilities and providing for hierarchies of abstraction,
• to improve support for re-use and incremental development by allowing further delaying of bindings through type-genericity and dynamic polymorphism,
• to preserve capabilities for synthesis and other forms of design analysis,
• to support hardware/software co-design through improved integration with programming languages (e.g., Ada),
• to support refinement of models through elaboration of components rather than through repartitioning, and
• to preserve correctness of existing models within the extended language.

Since SUAVE is an extension of the existing VHDL language, it is important that the extensions integrate well with all aspects of the existing language. In designing the SUAVE extensions, the design principals followed during the restandardization of VHDL that lead to the current language [15] were adopted in addition to those listed above. The goal was to preserve what Brooks refers to as the "conceptual integrity" of the language [11].

3. EXTENSIONS TO THE TYPE SYSTEM

Object-oriented languages support re-use and incremental development through the mechanism of inheritance. SUAVE extends the type system of VHDL by adopting the object-oriented features of Ada-95, including inheritance through type derivation and tagged types with extension on derivation.

3.1 Derived Types and Inheritance

For a type defined in a package, the operations (procedures and functions) defined in the package are called the *primitive operations* of the type. A new type can be defined as being *derived* from a parent type. In that case, the derived type inherits the set of values and the primitive operations of the parent type. An inherited operation can be overridden by defining a new operation with the same name but with operands of the derived type. Furthermore, additional primitive operations can be defined for the derived type. SUAVE adopts the Ada notation for defining a derived type, for example:

type event_count **is new** natural;

The derived type is distinct from, but related to, the parent type. Use of derived types helps avoid inadvertent mixing of conceptually different values, and thus improves the expressiveness of the language.

3.2 Tagged Types and Type Extension

As in Ada-95, a record type in SUAVE may include the reserved word **tagged** in its definition. Such a type is called a *tagged type*. An object of a tagged type includes a run-time tag that identifies the specific type used to create the object. The tag is used for dynamic dispatching, which is described below. A tagged record type may be *extended* by deriving a new type with a *record extension* containing additional record elements. This is the origin of the term "programming by extension," sometimes used to describe the Ada-95 approach. The derived type is also a tagged type that can be further extended. Since all elements in the parent type are also in the derived type, inherited operations of the parent type can be applied to objects of the extended type. However, any overriding or newly defined operations for the extended type can only be applied to the extended type (or its derivatives), since they may refer to the elements in the extension.

As an example, consider a type and operations representing an instruction set for a RISC CPU. All instructions have an opcode. ALU instructions additionally have fields for the source and destination register numbers. Thus an ALU instruction can be considered as an extension of a base instruction with just an opcode. This can be expressed in SUAVE by defining the following in a package:

```
type instruction is tagged record
        opcode : opcode_type;
    end record instruction;

function privileged ( instr : instruction; mode : protection_mode )
            return boolean;

procedure disassemble ( instr : instruction; file output : text );

type ALU_instruction is new instruction with record
        destination, source_1, source_2 : register_number;
    end record ALU_instruction;

procedure disassemble ( instr : ALU_instruction; file output : text );
```

The subprograms privileged and disassemble are primitive operations of instruction and are inherited by derived types. The type ALU_instruction is derived from instruction and has four elements: the opcode element inherited from instruction, and the three register number elements defined in the extension. A version of the function privileged is inherited from instruction with the instr parameter being of type ALU_instruction. The disassemble instruction defined for ALU_instruction overrides that inherited from instruction.

3.3 Abstract Types and Subprograms

An *abstract type* is a tagged type that is intended for use solely as the parent of some other derived type. Objects may not be declared to be of an abstract type. An *abstract subprogram* is one that has no body (and requires none), because it is intended to be overridden when inherited by a derived type. Abstract types and subprograms allow definition of types that include common properties and operations, but which must be refined by derivation of types that represent concrete objects.

As an illustration, consider refinement of the instruction type to represent memory reference instructions using displacement addressing mode. Such instructions include a base register number and an offset. The type for these instructions is declared abstract, since it is intended to be the parent type for load and store instruction types. More precisely,

```
type memory_instruction is abstract new instruction with record
        base : register_number
        offset : integer:
    end record memory_instruction;

function effective_address_of ( instr : memory_instruction ) return integer;

procedure perform_memory_transfer ( instr : memory_instruction ) is abstract;
```

The function effective_address_of is not abstract, since it can calculate the result using the data in a memory_instruction record. The function can be inherited "as is" by derived types. The procedure perform_memory_transfer, on the other hand, is declared abstract since the direction of transfer depends on whether a memory instruction is a load or a store. The derived types must provide overriding non-abstract implementations of this procedure. Examples are derived types for load and store instructions, as follows:

```
type load_instruction is new memory_instruction with record
        destination : reg_number;
    end record load_instruction;
procedure perform_memory_transfer ( instr : load_instruction );
procedure disassemble ( instr : load_instruction; file output : text );
type store_instruction is new memory_instruction with record
        source : reg_number;
    end record store_instruction;
procedure perform_memory_transfer ( instr : store_instruction );
procedure disassemble ( instr : store_instruction; file output : text );
```

Objects cannot be declared to be of type memory_instruction, but they can be declared to be of type load_instruction or store_instruction.

3.4   Class-Wide Types and Operations

One of the most important aspects of object-oriented programming is the use of *classes*. SUAVE adopts the Ada-95 mechanism of *class-wide types* to deal with classes. This contrasts with languages such as Simula [13], C++ [21]and Java [14] that introduce a special construct for classes. (See our paper that compares the two approaches [2].)

Class-wide types are denoted using the 'Class attribute. For a tagged type T, the *class-wide type* denoted T'Class is the union of T and all types derived directly or indirectly from T. The type T is called the *root* of the class-wide type. For example, the class-wide type instruction'class denotes the hierarchy of types rooted at instruction, and including ALU_instruction, memory_instruction, load_instruction and store_instruction.

An object of a class-wide type can have a value of any specific type in T'Class. Such an object is called *polymorphic*, meaning that it can take on values of different types during its lifetime. SUAVE allows constants, dynamically allocated variables and signals to be of a class-wide type. When an operation is applied to an object of a class-wide type, the tag of the value is used to determine the specific type, and thus to determine which primitive operation to invoke. This is called *dynamic dispatching*, or *late binding*, and is an essential aspect of object-oriented languages. As an example, consider the following signal declaration and application of an operation:

```
signal fetched_instruction : instruction'class;
disassemble ( fetched_instruction );
```

If the value of the signal is of type instruction, the version of disassemble for that type is invoked. However, if the value of the signal is of one of type load_instruction, the overriding version defined for load_instruction values is invoked. The choice is made dynamically at the time of the call.

While there are no primitive operations of a class-wide type, a subprogram may have a parameter of a class-wide type. Such a subprogram is called a *class-wide operation*. For example:

procedure execute (instr : instruction'class);

Since the parameter is polymorphic, dynamic dispatching may be required for operations on the parameter within the subprogram.

As a final example in this section, consider an instruction register that can jam a TRAP instruction in place of the store instruction. First, two constants are declared for the TRAP instruction and an undefined instruction:

constant halt_instruction : instruction := instruction'(opcode => op_halt);

constant undef_instruction : instruction := instruction'(opcode => op_undef);

Next, the entity is declared:

entity instruction_reg **is**
    port (load_enable : **in** bit; jam_halt : **in** bit;
            instr_in : **in** instruction'class; instr_out : **out** instruction'class);
    end entity instruction_reg;

The ports instr_in and instr_out are signals of a class-wide type and so may take on values of any of the types in the instruction hierarchy. A behavioral architecture body for the register is:

architecture behavioral **of** instruction_reg **is**
begin
    store : **process** (load_enable, jam_halt, instr_in) **is**
        type instruction_ptr **is access** instruction'class;
        variable stored_instruction : instruction_ptr := **new** undef_instruction;
    begin
        if jam_halt = '1' **then**
            deallocate (stored_instruction);
            stored_instruction := **new** halt_instruction;
        elsif load_enable = '1' **then**
            deallocate (stored_instruction);
            stored_instruction := **new** instr_in;
        end if;
        instr_out <= stored_instr.**all**;
    end process store;
end architecture behavioral;

The process implements the register storage using the local variable stored_instruction. Since a variable cannot be of a class-wide type, stored_instruction is defined as an access value, pointing to a dynamically allocated object of type instruction'class. It is initialized to the undefined instruction. When a HALT instruction is to be jammed, a new instruction object initialized to the halt instruction value is allocated. Similarly, when an input instruction is to be stored, a new instruction object of the corresponding specific type is allocated and initialized to the input instruction. The designated instruction object is assigned as the output of the register.

4. EXTENSIONS FOR ENCAPSULATION

A data type in VHDL is characterized by a set of values, specified by a type definition, and a set of operations. An *abstract data type* (ADT) is one in which the concrete details of the type definition are hidden from the user of the ADT. The user may only use the operations of the ADT to manipulate values. ADTs are important tools for managing complexity in a large design.

VHDL currently includes the *package* feature, which can be used to define an ADT. The concrete type and associated operations are declared in the package declaration, and the implementations of the operations are declared in the package body. While this approach allows the implementation details of the operations to be hidden from the ADT user, it exposes the details of the concrete type. A user may inadvertently (or deliberately) modify values of the concrete type directly, rather than by using the provided operations. This can potentially place the ADT value in an inconsistent state. It also reduces the maintainability of the design.

SUAVE extends the type system and package feature of VHDL to provide secure encapsulation of information in an ADT. It adopts the mechanisms of private types and private parts in packages from Ada-95. This meets one of the design objectives for SUAVE: to improve encapsulation and information hiding.

As a first step, the use of packages is generalized by allowing them to be declared as part of most declarative regions in a model, not just as library units. SUAVE allows a package declaration and body to be declared in an entity declaration, an architecture body, a block statement, a generate statement, a process statement, and a subprogram body. Thus, the concept of a package is changed from that of a "heavy-weight" library-level unit to that of a "lightweight" declarative item. This is important, since packages are used to declare types and operations defining classes, as well as instances of generic packages (see Ashenden *et al* [5]).

4.1 Private Parts and Private Types

The second extension of the package feature is to allow a package declaration to be divided into a *visible part* and a *private part*, as follows:

package *name* **is**
   ...   -- *visible part*
private
   ...   -- *private part*
end package *name*;

Items declared in the visible part are exported and may be referred to by users of the package. Items declared in the private part, on the other hand, are not visible outside the package. When using a package to define an ADT, the type is declared as a *private type* in the visible part of the package, along with the specifications of the primitive operations of the type. A private type declaration only provides the name of the type. The concrete details of the type are declared separately in the private part of the package.

As an example, the following package defines an ADT for complex numbers:

```
package complex_numbers is
    type complex is private;
    constant i : complex;
    function cartesian_complex ( re, im : real ) return complex;
    function re ( C : complex ) return real;
    function im ( C : complex ) return real;
    function polar_complex ( r, theta : real ) return complex;
    function "abs" ( C : complex ) return real;
    function arg ( C : complex ) return real;
    function "+" ( L, R : complex ) return complex;
    function "-" ( L, R : complex ) return complex;
    function "*" ( L, R : complex ) return complex;
    function "/" ( L, R : complex ) return complex;
private
    type complex is record
            r, theta : real;
        end record complex;
end package complex_numbers;
```

A user of this package can declare objects of type complex and invoke operations on complex numbers, for example:

```
signal x, y, z : complex := cartesian_complex(0.0, 0.0);
. . .
z <= x * y after 20 ns;
```

However, the fact that complex numbers are represented in polar form is hidden. Indeed, the representation may be changed without requiring changes to the user's code.

4.2   Private Extensions

SUAVE adopts the Ada-95 mechanisms for integrating encapsulation with inheritance. A private type can be declared to be tagged, indicating that it can be used as the parent of a derived type. The concrete details remain hidden in the private part of the package. A tagged private type can also be declared abstract if it should not be directly instantiated. For example, a network packet at the media-access level of a protocol suite might be declared as follows:

```
package MAC_level is
    type MAC_packet is abstract tagged private;
    . . .
private
```

```
    type MAC_packet is tagged record
        ...
        end record MAC_packet;
    end package MAC_level;
```

A tagged private type can be extended using type derivation, as described in Section3. However, for the derived type to take on the form of a secure ADT, it should be declared as a *private extension*. This allows the details of the extension to be encapsulated. For example, the network packet type defined above may be extended with payload information to form a network-level packet:

```
    package network_level is
        type network_packet is new MAC_packet with private;
        ...
    private
        type network_packet is new MAC_packet with record
            ...
        end record network_packet;
    end package network_level;
```

A user of this package knows that a network-level packet is derived from a MAC-level packet, and thus inherits all of the operation applicable to a MAC-level packet. The concrete details of both types, however, remain hidden.

5. EXTENSION OF GENERICS

VHDL currently allows an entity declaration, a component declaration or a block statement to include a *generic clause*, which defines formal generic constants for the unit. Generic constants are typically used to specify timing and other operational parameters and to specify index bounds for array ports. When a unit with a generic clause is instantiated, a *generic map aspect* is used to associate actual values with the formal generic constants. The actuals are constants whose values are used in place of the formal generic constants for this instance. Association of actuals with formal generics occurs when the instance is elaborated prior to simulation or synthesis.

5.1 Overview of Extensions of Generics

One of the main aspects that constrains re-use of the multiplexer entity described above is that it can only be instantiated to deal with bit-vector values. A more re-usable multiplexer entity would be instantiable for arbitrary types. Thus, it is desirable to be able to specify the data type as a formal generic. In many cases, this is feasible, since the implementation of a unit does not depend on the details of any particular type. For example, a behavioral implementation of a multiplexer simply involves assigning values from input to output, irrespective of the type of the values. A given multiplexer instance, however, should only be allowed to deal with values of one particular type, namely the type of the signals connected to its data ports. This restriction is in conformance with the strong-typing philosophy of VHDL.

SUAVE extends the generic clause feature of VHDL by allowing specification of formal generic types. A unit may use a formal type to define ports and other objects in its implementation. When the unit is instantiated, an actual type is associated with the formal type for that instance. The association occurs when the instance is elaborated.

The particular mechanism for specifying formal types is modeled on the corresponding mechanism in Ada-95 {ISO/IEC, 1995 #4}, but is adapted to integrate cleanly with the existing generic mechanism in VHDL. A formal generic type is specified in the following form in a generic clause:

type *identifier* **is** *interface_type_definition*

For example, a multiplexer entity might include a formal generic type for the data to be handled.

```
entity mux is
    generic ( type data_type is private );
    port ( sel : in bit; d0, d1 : in data_type; d_out : out data_type );
end entity mux;
```

SUAVE allows a number of different classes of type definition, each restricting the actual type that can be associated when the unit is instantiated, as shown in Table 1. (Further refinements to the first three classes are described in the SUAVE report [6].) The implementation of a unit can make use of the knowledge about the formal type afforded by the definition. For example, it may use arithmetic operations on a formal integer type, or indexing on a formal array type.

The actual value to be associated with a formal type is specified as a type name in the generic map aspect. For example, the generic multiplexer described above might be instantiated for integer data types as follows:

```
int_mux : entity work.mux(behavioral)
    generic map ( data_type => integer )
    port map ( . . . );
```

SUAVE further extends VHDL by allowing package declarations and subprogram specifications to include generic clauses, enabling definition of template packages and subprograms that can be re-used with different type bindings. This feature combines with the object-oriented extensions in SUAVE [7] to provide means of defining generic abstract data types in a type-secure way.

A generic package includes a generic clause before the declarations in the package, for example:

```
package float_ops is
    generic ( type float_type is range <>.<> );
    . . .
end package float_ops;
```

Table 1: Classes of formal generic types in SUAVE.

Formal type	Restrictions on actual type
private	actual can be any type that allows assignment
new *type_mark*	actual must be derived from the specific type (see [7])
new *type_mark* **with private**	actual must be derived from the specific tagged type (see [7])
(<>)	actual must be a discrete type
range <>	actual must be an integer type
units <>	actual must be a physical type
range <>.<>	actual must be a floating-point type
array (*index_type*) **of** *element_type*	actual must be an array type with the specified index and element types
access *subtype*	actual must be an access type with the specified designated type
file of *type_mark*	actual must be a file type with the specified element type

A generic package such as this cannot be used directly. Instead, it must be instantiated and actual generics associated with the formal generics, for example:

```
type amplitude is range -10.0 to +10.0;
package amplitude_ops is new float_ops
    generic map ( float_type => amplitude );
```

Note that the instance is a package declared within an enclosing declarative region. SUAVE generalizes the use of packages by allowing them to be declared in inner regions, rather than just as library units. This generalization is also related to the use of packages in the object-oriented extensions, and is discussed further in that context [7].

A generic subprogram includes a generic clause before the parameter list, analogous to the way in which an entity includes the generic clause before the port list, for example:

```
procedure swap
    generic ( type data_type is private )
    ( a, b : inout data_type ) is
    variable temp : data_type;
begin
    temp := a;  a := b;  b := temp;
end procedure swap;
```

A generic subprogram cannot be called directly, but must be instantiated first, for example:

```
procedure swap_times is new swap
    generic map ( data_type => time );
```

This declares a procedure with two parameters of type time. A call to the procedure includes a normal actual parameter list, for example:

```
swap ( old_time, new_time );
```

6.   CONCLUSION

In this paper we have described the SUAVE extensions to VHDL to improve its support for modeling at all levels of abstraction. We have presented the features that provide object-orientation as a combination of improved abstraction, encapsulation and inheritance mechanisms, and the genericity features that improve support for re-use. Most of the features are drawn from Ada-95 and are adapted to integrate with modeling features that are specific to VHDL. Drawing on Ada is appropriate, since VHDL was originally strongly influenced by Ada. In a sense, SUAVE is an evolution of VHDL that parallels the evolution from Ada-83 to Ada-95.

SUAVE improves modeling support by generalizing and extending existing mechanisms, rather than by adding whole new features. In particular, SUAVE avoids replication of the abstraction & encapsulation mechanisms already provided by the package feature. Adding a separate class feature, as proposed in Objective VHDL [19], for example, replicates many aspects of packages and so complicates a designer's choice of expression of design intent.

Space considerations preclude a more detailed definition of the features added in SUAVE. In particular, we have omitted description of formal generic subprograms and packages. The interested reader can find a more complete description in the SUAVE report [6]. Work is now in progress to implement the extensions within the framework of the SAVANT project [18].

7.   REFERENCES

1.   P. J. Ashenden, *The Designer's Guide to VHDL*. San Francisco, CA: Morgan Kaufmann, 1996.

2.   P. J. Ashenden and P. A. Wilsey, *A Comparison of Alternative Extensions for Data Modeling in VHDL*, Dept. Computer Science, University of Adelaide, Technical Report TR-02/97, ftp://ftp.cs.adelaide.edu.au/pub/VHDL/TR-data-modeling.ps, 1997.

3.   P. J. Ashenden and P. A. Wilsey, "Considerations on Object-Oriented Extensions to VHDL," *Proceedings of VHDL International Users Forum Spring 1997 Conference*, Santa Clara, CA, pp. 109–118, 1997.

4.   P. J. Ashenden and P. A. Wilsey, *Principles for Language Extension to VHDL to Support High-Level Modeling*. Dept. Computer Science, University of Adelaide, Technical Report TR-03/97, ftp://ftp.cs.adelaide.edu.au/pub/VHDL/TR-principles.ps, 1997.

5.   P. J. Ashenden, P. A. Wilsey, and D. E. Martin, "Reuse Through Genericity in SUAVE," *Proceedings of VHDL International Users Forum Fall 1997 Conference*, Arlington, VA, pp. 170–177, 1997.

6.   P. J. Ashenden, P. A. Wilsey, and D. E. Martin, *SUAVE: A Proposal for Extensions to VHDL for High-Level Modeling*, Dept. Computer Science, University of Adelaide, Technical Report TR-97-07, ftp://ftp.cs.adelaide.edu.au/pub/VHDL/TR-extensions.pdf, 1997.

7.   P. J. Ashenden, P. A. Wilsey, and D. E. Martin, ""SUAVE: Painless Extension for an Object-Oriented VHDL," *Proceedings of VHDL International Users Forum Fall 1997 Conference*, Arlington, VA, pp. 60–67, 1997.

8.   J. Barnes, Ed. *Ada 95 Rationale*, Lecture Notes in Computer Science, vol. 1247. Berlin, Germany: Springer-Verlag, 1997.

9.   J. Benzakki and B. Djaffri, "Object Oriented Extensions to VHDL: the LaMI Proposal," *Proceedings of Conference on Hardware Description Languages '97*, Toledo, Spain, pp. 334–347, 1997.

10.   G. Booch, *Object-Oriented Analysis and Design with Applications*. Redwood City, CA: Benjamin/Cummins, 1994.

11.   F. P. Brooks, Jr., *The Mythical Man-Month, Anniversary ed.* Reading, MA: Addison-Wesley, 1995.

12.   D. Cabanis and S. Medhat, "Classification-Orientation for VHDL: A Specification," *Proceedings of VHDL International Users Forum Spring '96 Conference*, Santa Clara, CA, pp. 265–274, 1996.

13.   O. J. Dahl and K. Nygaard, "Simula: An Algol Based Simulation Language," *Communications of the ACM*, vol. 9, no. 9, pp. 671–678, 1966.

14.   J. Gosling, B. Joy, and G. L. Steele, *The Java Language Specification*. Reading, MA: Addison-Wesley, 1996.

15.   IEEE, *Standard VHDL Language Reference Manual*. Standard 1076-1993, New York, NY: IEEE, 1993.

16.   ISO/IEC, *Ada 95 Reference Manual*. International Standard ISO/IEC 8652:1995 (E), Berlin, Germany: Springer-Verlag, 1995.

17.   M. T. Mills, *Proposed Object Oriented Programming (OOP) Enhancements to the Very High Speed Integrated Circuits (VHSIC) Hardware Description Language (VHDL)*, Wright Laboratory, Dayton, OH. Tech. Report WL-TR-5025, 1993.

18.   MTL Systems Inc., *Standard Analyzer of VHDL Applications for Next-generation Technology (SAVANT)*. MTL Systems, Inc, http://www.mtl.com/projects/savant/, 1996.

19.   M. Radetzki, W. Putzke, W. Nebel, S. Maginot, J.-M. Bergé, and A.-M. Tagant, "VHDL Language Extensions to Support Abstraction and Re-Use," *Proceedings of Workshop on Libraries, Component Modeling, and Quality Assurance*, Toledo, Spain, 1997.

20.   G. Schumacher and W. Nebel, "Inheritance Concept for Signals in Object-Oriented Extensions to VHDL," *Proceedings of Euro-DAC '95 with Euro-VHDL '95*, Brighton, UK, pp. 428–435, 1995.

21.   B. Stroustrup, *The C++ Programming Language*. Reading, MA: Addison-Wesley, 1986.

22.   S. Swamy, A. Molin, and B. Covnot, "OO-VHDL: Object-Oriented Extensions to VHDL," *IEEE Computer*, vol. 28, no. 10, pp. 18–26, 1995.

23.   A. Taivalsaari, "On the Notion of Inheritance," *ACM Computing Surveys*, vol. 28, no. 3, pp. 438–479, 1996.

24.   P. Wegner, "Dimensions of Object-Based Language Design," *ACM SIGPLAN Notices*, vol. 22, no. 12, *Proceedings of OOPSLA '87*, pp. 168–182, 1987.

25.   J. C. Willis, S. A. Bailey, and R. Newschutz, "A Proposal for Minimally Extending VHDL to Achieve Data Encapsulation Late Binding and Multiple Inheritance," *Proceedings of VHDL International Users Forum Fall '94 Conference*, McLean, VA, pp. 5.31–5.38, 1994.

DIGITAL CIRCUIT DESIGN WITH OBJECTIVE VHDL

Martin Radetzki, Wolfgang Nebel
OFFIS Research Center, Escherweg 2, 26121 Oldenburg, Germany

Abstract. Objective VHDL is an extension of the hardware description language VHDL. It provides the constructs necessary to design hardware in an object-oriented way. Thereby, proven software engineering concepts for abstraction and reuse can be transferred to the hardware world. Moreover, the use of Objective VHDL facilitates the use of object-orientation as a design methodology that embraces both the hardware and software parts of an embedded system, independent of the fact that a different object-oriented language may be used for software. In addition to such specification techniques, we outline synthesis steps to generate high-level VHDL from an object-oriented system level model, enabling the use of commercial behavioral synthesis tools and simulators. All these concepts are demonstrated with an example used consistently in this contribution.

1.   INTRODUCTION

Hardware synthesis from object-oriented programming languages such as C++ and Java has recently attracted a lot of interest from researchers and users. Initial specification and evaluation of algorithms are frequently performed in these languages for performance reasons, and direct synthesis can help to avoid the effort of re-coding them in an HDL. The embedded systems context, with most effort going into software development, gives further impact to the application of software-like techniques for hardware design.

We investigate several approaches for describing and synthesizing hardware using object-orientation. A lot of object-oriented HDLs, mostly VHDL extensions, have been proposed, but there is little work on their synthesis. Many programming language approaches, such as Ocapi [1] and SystemC [2], provide an object-oriented specification framework, but do not synthesize the object-oriented constructs themselves. Others, e.g. Matisse [3] and SystemC++ [4], are promising in that they could provide more synthesis support for objects.

71

J. Mermet (ed.), Electronic Chips & Systems Design Languages, 71–84.
© 2001 *Kluwer Academic Publishers.*

In the third section, we present a more general, language-independent conceptual framework for the system-level synthesis of not only objects, but also their communication in a concurrent context. The language Objective VHDL [5], presented in section four, has been designed to reflect this framework and its restrictions necessary for a hardware implementation of object-oriented models. In the fifth section, the main challenges and synthesis concepts are presented with an example. As shown in section six, high-level behavioral VHDL code can be generated and fed into commercial synthesis tools in order to obtain a gate-level implementation of the object system. Simulation runs at all levels of abstraction demonstrate that the synthesis concepts provide a correct hardware implementation of an object-oriented system description.

2. RELATED WORK

The related work on hardware description employing object-oriented techniques can be classified into three groups: object-oriented extensions of hardware description languages (mostly VHDL), programming language based hardware description frameworks, and hardware synthesis of object-oriented programming language constructs.

Since the early 1990s, researchers have produced a considerable amount of object-oriented VHDL dialects such as VHDL_OBJ [6], VHDL++ [7], Vista [8] and LaMI [9]. Nowadays only Ashenden's SUAVE [10], Schumacher's OO-VHDL [11], and Objective VHDL [5] are continued. Like most earlier approaches, SUAVE targets primarily at simulation. While large parts of the language may be synthesizable, no respective techniques have been published up to now. Different to that, Schumacher presents a translation from OO-VHDL into VHDL, claiming the synthesizability of the generated code. However, the translation makes use of records and subprograms that are not well supported by current VHDL synthesis tools. Moreover, the suggested coding style for the synchronization of concurrent requests to an object is innovative in that it avoids issues known as inheritance anomaly [12], but it is definitively not synthesizable. The next section of this contribution describes the synthesis of objects and their communication independent of an implementation language. Still, such a language is needed, and we opt for the use of Objective VHDL because this language has been designed to ease synthesis by incorporating some hardware-related restrictions and hardware-specific semantics.

The class of programming language based hardware description environments includes SystemC [2] and Ocapi [13]. Both have in common that the object-oriented features of a programming language are not directly used for describing hardware. Rather, they are used to provide class libraries to augment the programming language with features such as events, reactivity, time, and concurrency that are essential for hardware description but not incorporated in the language itself. Hence, synthesis—if addressed at all—does not transform the object-oriented constructs themselves into hardware.

This is different with the third class, comprising Matisse [3], SystemC++ [4] (not to be mixed up with SystemC), and JavaSynth [14]. Matisse maps a C++ model onto a specific hardware architecture, storing all object data in a central memory and implementing functionality, including data memory access, as an application specific digital circuit. The storage model is similar to that chosen by compilers of object-oriented programming languages, enabling the dynamic allocation and deallocation of objects. SystemC++, on the other hand, chooses to implement objects in separate function units with individual storage. Hence, all objects can operate concurrently whereas a global memory limits concurrency by the number of its ports. The dynamic creation and destruction of objects is supported by a run-time binding of objects to the computation resources. As with dynamic memory allocation, this implies not only a hardware overhead for implementing the required run-time system. Even worse, dynamic allocation makes system behavior less predictable, less suitable for formal verification, and can lead to deadlocks when allocation goes beyond the available resources. A similar argumentation holds for the use of references, e.g. in the JavaSynth approach. Since it is not always possible to determine statically to which object a reference points, dynamic mechanisms may have to be implemented in hardware for this purpose.

3. LANGUAGE-INDEPENDENT CHARACTERIZATION OF SYNTHESIZABLE OBJECT SYSTEMS

The approaches mentioned above have contributed to the investigation of the synthesis of object-oriented features present in popular programming languages. The dynamic nature of object-oriented concepts found in many languages deviates from the traditional understanding that a hardware description must be static. Some researchers define hardware target architectures with dynamic properties, e.g. dynamic memory allocation or dynamic resource management, that allow to directly implement the source language. Others limit the use of dynamic features to an extent that little object-orientation remains, or suggest the use of object-orientation only to provide a framework for hardware description.

Our approach is based on the definition of a meta model of object-orientation which is suitable for translation into traditional, statically allocated, bound, and scheduled hardware resources. The model is independent of a source language. We can either define a language that reflects our model, or impose limitations and coding styles on the use of a programming language to restrict dynamics to the degree supported for synthesis. In the following, we outline the properties of the meta model.

3.1   Object state

Each object has its individual state which may vary over time. The state space is defined by a class declaration. A class can be understood as a template for the creation of similar objects. A class defines the state space of each of its objects as a collection of so-called attributes. This term, to be told apart from VHDL attributes, means the definition of a variable data field with a given type. This type defines the values which can be stored in the attribute, i.e., the attribute's state space. The state space of the object is the product of its attributes' state spaces.

A derived class extends the definition of the parent class from which it is derived. The derived class inherits all attributes from the parent class and may declare additional ones. Thereby it defines a new state space which is the parent's state space extended for the additional attributes.

The knowledge of the state space allows us to compute the number of bits which must be allocated in a memory or register to store an object's state. There is so far no fundamental difference compared with software. We only have to demand that all attributes' types be synthesizable. This means, in particular, that they must not be pointers.

3.2   State transitions

In addition to the state space, classes also define methods (operations, services) which allow to modify an object's state. In the strict sense of object-orientation, there should be no other way to perform a state transition of an object than by invocation of its methods.

A method is similar to a subprogram (procedure or function). It can have input and output parameters, and it has access to the object's state. A typical method invocation is as follows: The method receives a number of input values via its input parameters, reads and modifies the object's state, and returns a number of output values via its output parameters. In some sense, an object can therefore be compared to a finite state machine (FSM). It has a state, inputs and outputs defined by its methods' parameters, and state transition and output functions as defined by the methods' behavior. An FSM can be synthesized into hardware, and indeed the circuit structure that implements an object (see Section 5.1) reminds of the typical FSM implementation with state register, output and next-state logic, and feedback. However, the FSM model is only of a conceptual nature because:

• The state space of an object is typically too large to apply state minimization and encoding optimization algorithms.

• The description of the methods' functionality is at the algorithmic level, using imperative statements and control structures such as assignments, conditional statements, and loops. Method execution may therefore (have to) take more than a single clock cycle, whereas FSM state transition functions are typically combinational.

Later in this paper, we will devise techniques for the implementation of the complex state space and the methods. The application of high-level synthesis (HLS) will allow to synthesize methods as long as their code is written in a synthesizable style. This is not a severe limitation compared to object-orientation in software. All the typical statements, even loops with dynamic boundaries, can be handled well by HLS.

3.3   Object lifetime

In most programming languages, object allocation is dynamic. C++ and Java provide the **new** operator to create an object. Storage for the object's state is allocated in heap memory. Similarly, objects may be destroyed while the system is at operation. Then, the memory which has been occupied by its state is freed and made available, possibly with the help of a garbage collector, for the next allocation.

Of course, memory resources are limited. Object allocation may fail if the number of objects created exceeds the available resources. Moreover, it is difficult to assess the maximum number of resources needed if they are dynamically allocated. Therefore, we only allow static object allocation in our meta model of hardware object-orientation. Staticness is easily ensured when using signals and variables in Objective VHDL. For languages such as C++ or Java, we can demand that objects be only allocated, e.g., in constructors, and determine these allocations in an elaboration phase similar to VHDL's elaboration.

3.4   Communication between objects

Up to now we have only considered single objects. In any real-world system, we can expect to find several, communicating objects. The object-oriented way of communication is called message passing. By passing a message to a target object (server), a client object requests the execution of a method from the server. The message includes identification of the requested method and values for its input parameters. The message passing mechanism may also take care of transmitting the values of output parameters back to the client. In sequential software, this is usually implemented as a subprogram call. Considering the concurrent nature of hardware, we will keep the more general notion of message passing.

To pass a message to a server, it is first necessary to identify the server. We will say that the client must have a *channel* to the server. In object-oriented software, a channel is established if the client owns a reference (pointer) to the server. References may be passed dynamically from object to object, e.g., as method parameters. Thus, channels can be established dynamically. We find a similar situation in distributed object-oriented software systems, where references are usually replaced by handles (an abstraction of memory addresses), while the dynamic nature stays the same.

Targeting hardware synthesis, we require all communication channels to be static. This is possible since all objects are created statically in our model. The staticness of channels can be enforced in Objective VHDL by using signals which are all determined during elaboration. The signals can be used to describe the channels, but they can also serve as objects. In the latter case, signal references passed through the system would serve as channels. Also these references are statically elaborated.

3.5 Request arbitration

In a concurrent hardware model, a server object may receive several messages (i.e., method execution requests) at the same time from several concurrent clients. Each method has access to the object's state, which is a shared resource. A concurrent execution of methods could therefore lead to inconsistencies due to interleaving of concurrent write and read operations. Hence, method execution must be sequentialized, and each method must be executed atomically. This property is implicit in our model. It must be implemented by a dedicated arbitration unit which can be instantiated automatically by an object synthesis tool (see Section 5.3).

Moreover, an object may at times be unable to provide a service. For instance, an empty buffer cannot provide a GET service which takes an element out of the buffer. In this case, the GET request would have to be blocked until a new element has been put into the buffer from another, concurrent client. A modeling guideline that allows to define such so-called *condition synchronization* has been devised in [11]. It must be carefully designed to avoid the hazard of code becoming invalid during inheritance, known as *inheritance anomaly*. From the model, boolean-valued *guard* functions can be extracted. A guard belongs to a method. The guard is true if and only if the method can be safely executed by an object in its current state. This must be taken into account by the arbitration unit.

3.6 Example: FIFO buffer class

Figure 1 shows the UML diagram and the implementation concept of a first-in-first-out (FIFO) buffer class. It is parameterized with its maximal number of entries, SIZE. Entries of type T_ELEMENT are stored in the attribute STORAGE which is organized in a circular manner. The attribute FIRST stores the index position within STORAGE where the oldest element is stored, while NXT points to the next free position in the buffer. If FIRST equals NXT, the buffer is either empty or full. This is distinguished with the attribute EMPTY. Together, the attributes define the state space of a FIFO object.

The methods IS_EMPTY and IS_FULL report the status of the buffer. They can always be invoked; their guard could be said to be TRUE. Method PUT allows to put an element into the buffer. It can be invoked only if the buffer is not full. Therefore, it is guarded by the expression not IS_FULL. Method GET takes an element out of the buffer. Its guard is the boolean expression not IS_EMPTY. A listing of the methods is provided below. We will continue the example in the following sections with several clients connected by channels to a FIFO object.

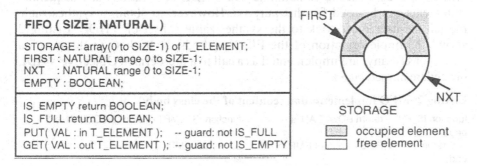

Figure 1. Class FIFO

4. SYSTEM DESCRIPTION WITH OBJECTIVE VHDL

Objective VHDL provides a new kind of type, a class type similar to the C++ class construct, in addition to VHDL's type system. The class type provides a generic clause to allow its parameterization, and auxiliary declarations of, e.g., types used only by the class, can be made inside the class as shown in listing 1.

Listing 1: FIFO declaration as a class type

```
type FIFO is class
    generic( SIZE : NATURAL );
    type STORAGE_ARRAY is array(0 to SIZE-1) of T_ELEMENT;
    class attribute STORAGE : STORAGE_ARRAY;
    class attribute FIRST, NXT : NATURAL range 0 to SIZE-1;
    class attribute EMPTY : BOOLEAN;
    for variable
        function IS_EMPTY return BOOLEAN;
        function IS_FULL return BOOLEAN:
        procedure PUT( VAL : in T_ELEMENT );
        procedure GET( VAL : out T_ELEMENT );
    end for;
end class FIFO;
```

The keywords **class attribute** allow to declare object-oriented attributes (not to be mixed up with VHDL's attributes) that constitute the state space of each object of the class. Finally, methods can be declared as subprograms (functions and procedures) inside the class. Since an object can either be a signal or a variable (in some cases also a constant), and since signal and variable semantics are different, methods may have to be implemented in different ways for signals and variables or may only be available for one of these alternatives. This can be declared by including method in a **for signal** or **for variable** block.

The implementation of methods is separated from the class type declaration in a class body with the same name. All VHDL declarations and sequential statements can be used for this purpose. However, to obtain a synthesizable model, the user must stick to the synthesizable subset of VHDL. Listing 2 shows the implementation of the FIFO methods. Note that synchronization conditions (if any) are implemented as a call to a pre-defined Objective VHDL procedure named guard.

Listing 2: FIFO implementation (content of the class body)

```
function IS_FULL return BOOLEAN is        function IS_EMPTY return BOOLEAN is
begin                                     begin
    return NXT = FIRST and not EMPTY;         return EMPTY;
end;                                      end;

procedure PUT(VALUE : in T_ELEMENT) is    procedure GET(VALUE : out T_ELEMENT) is
begin                                     begin
    guard( not IS_FULL );                     guard( not IS_EMPTY );
    assert not IS_FULL                        assert not IS_EMPTY ...
        report "FIFO : buffer overflow"       VALUE := STORAGE( FIRST );
        severity FAILURE;                     FIRST := (FIRST + 1) mod SIZE;
    STORAGE( NXT ) := VALUE;                  if FIRST = NXT then EMPTY := TRUE;
    NXT := (NXT + 1) mod SIZE;                else EMPTY := FALSE;
    EMPTY := FALSE;                           end if;
end;                                      end;
```

In addition to class types, there exist derived class types to provide inheritance, a reuse and extension feature a fully object-oriented language must provide. Moreover, polymorphism of objects, i.e. the uniform handling of objects that are related by inheritance, is available in Objective VHDL. While not discussed in the article at hand, the interested reader can find more information in [15].

Class types can be used to declare objects as signals or variables. Thanks to the language design of VHDL, all these objects are static and can be determined in an elaboration step. To ensure synthesizability and avoid dynamic mechanisms, only the use of references must be avoided by forbidding the use of access types (as in the VHDL synthesis subset).

An object, object1, can request a service (invoke a method) of another object, object2, by using the notation object2.method(parameters) as in C++. If such a method invocation is found in object1, we recognize a channel from object1 to object2. The need for arbitration arises if further objects operating concurrently with object1 have a channel to object2.

5.   CIRCUIT STRUCTURE TO BE SYNTHESIZED

In the following we show a digital circuit implementation of an object's state storage and methods, of channels for communication between objects, and of concurrent request arbitration.

5.1   Implementation of objects

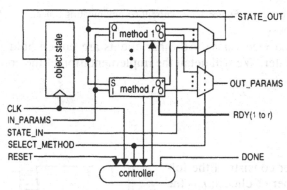

Figure 2. Object circuit structure

Figure 2 shows the structure of a circuit that implements an object. A register or memory large enough to store the object's state must be allocated. The state, as well as the input parameters, are being fed into data path circuits implementing the methods. The next state computed by the method to be executed is selected by a multiplexer under the control of the input SELECT_METHOD and is fed back into the state storage. Similarly, the correct output parameters are selected.

Since only a single method can be executed at a time, hardware resources can be shared among the method circuits. A controller is responsible for the execution of methods that take longer than a clock cycle, and reports their completion by a DONE signal.

The RDY output is another status signal. It has as many bits as there are methods. The value of RDY(i) is determined by the guard expression of the i-th method, which is computed as part of the method: RDY(i) = '1' if and only if the server's current state allows to execute the i-th method.

5.2   Implementation of channels

A static channel between a client and a server object is implemented by a set of wires. Some wires go from the client to the server. These wires are used to communicate from the client to the server's SELECT_METHOD input a binary code that identifies the method to be executed. They also transport the values of the input parameters to the server's IN_PARAMS input. Another set of wires goes from the server to the client. Over these wires, the DONE signal and values of the method's output parameters (if any) are returned to the client as soon as the method has been executed by the server.

Figure 3 shows a situation in which a server (the FIFO object) has *several* clients. Two clients, Producer1 and Producer2, issue PUT request to the FIFO buffer. A third client issues GET requests. The server can service only one of these requests at a time. Therefore, we introduce an additional block, the arbiter. The arbiter chooses to accept a request from one of the clients and connects

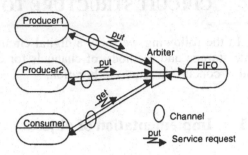

Figure 3. Concurrent communicating objects

this client with the server (and vice versa). All other clients are put on hold. Their request may be served later. We will detail the implementation of the arbiter in the following section.

5.3 Arbitration

The interface of the arbiter consists of the following signals (c is the number of clients; r is the number of different methods of the server):

- C_REQ(1 to c): requests received from the clients.
- S_RDY(1 to r): RDY signals from the server.
- S_REQ: the request forwarded to the server.
- C_IN(1 to c): input parameters from the clients.
- S_IN: input parameters forwarded to the server.
- S_DONE: DONE signal received from the server.
- C_DONE(1 to C): DONE signal to the clients.
- S_OUT: output parameters from the server.
- C_OUT(1 to C): output parameters to the clients.

The arbiter (Figure 4 shows a version synthesized from an RTL description) passes the clients' requests and the server status to a scheduler (the component at the left of Fig. 4). The scheduler re-

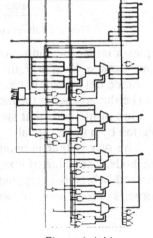

Figure 4. Arbiter

turns the number, $i \in \{1, ..., c\}$, of the client whose request shall be granted. All the arbiter has to do is to multiplex the request and input parameters of the i-th client to the server, and to de-multiplex the server's response to the i-th client while providing a stable DONE signal to the other clients. A special situation is if no request can be granted. The scheduler may indicate this case by returning $i = 0$, and the arbiter must send the server a no-operation request in this case.

The scheduler receives the signals C_REQ(1 to c), S_RDY(1 to r), and S_DONE from the arbiter and returns via an output, GRANT, a stable value of i as described above. Different schedulers may implement different scheduling policy. A scheduler may have an internal state. This allows, e.g., to memorize the client whose request has been granted last, so as to implement fair scheduling.

Figure 5 shows a scheduler circuit which has been synthesized from an RTL description implementing a static priority scheduling policy which serves client no. 1 first if it has a request. The RTL code is parameterized with the number of clients and the number of methods provided by the server. The particular instance shown to the right can handle up to four clients and four

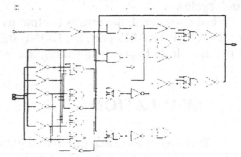

Figure 5. Scheduler

different methods. It can be used with the above FIFO example.

Other scheduling policies have as well been implemented; for instance, a round-robin scheme which serves the clients in a circular order.

6. VHDL CODE GENERATION

On the top level, VHDL code is generated as follows:
• A class is transformed into an entity-architecture pair. The entity interface consists of ports corresponding to the inputs and outputs shown in figure 2. Different architectures can be used to represent different optimizations for individual objects.
• An object corresponds to the instantiation of the entity-architecture pair that represents a particular optimization of its class.
• A channel is implemented by signals which are connected to the respective ports of the client and server object instances.
• Whenever more than one channel goes to an object, an arbitration component is instantiated and connected to the object. All clients are connected to the arbiter, not to the object itself.

While we cannot go into all the details (these are documented in [15]), we outline how an architecture is generated from a class. This aspect is particularly important because the architecture must follow the coding style acceptable by a HLS tool (namely Synopsys BC) to enable further synthesis using a VHDL-based design flow:
• A single process with nested reset and main loops is generated.
• In the reset loop, initial values are assigned to the class attributes (which, in turn, are translated into process variables).
• In the main loop, a request is awaited by a synthesizable busy loop. When a request is detected, the DONE signal is set '0' to indicate that the object cannot accept another request.

- A case statement inspects the request identifier and executes the code of the corresponding method. The HLS tool may schedule this operation in multiple clock cycles.
- When completed, the results (output parameters of a method) are assigned to the OUT_PARAMS port, the DONE signal is set to '1', and a new iteration of the main loop begins by awaiting the next request.

7. SIMULATION

A FIFO object for a maximum of eight entries, synthesized into a structure as shown in Fig. 2, has been simulated together with three clients as shown in Figure 3. Arbitration and fair round-robin scheduling were performed by synthesized circuits. The simulation result is displayed in Figure 6. Both producers (cf. Figure 3) issue PUT messages which are encoded as value 1 on the C1_REQ and C2_REQ signals. The consumer sends GET requests, encoded by the value 2 on the C3_REQ signal. A low pulse on the Ci_DONE signal tells the client no. i that its request is being served. This client acknowledges by sending a no-operation request (value 0) before issuing the next PUT. The signals prefixed with an S_ belong to the server. S_REQ is the request sent to the server by the arbiter, and S_DONE is the DONE signal from the server. The D_IN and D_OUT signals carry the input parameter value of PUT and the output of GET, respectively. The first producer (C1) always puts the token '1' into the buffer, and the second one (C2) the token '2'.

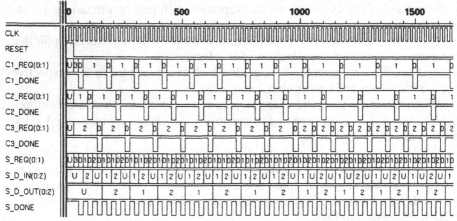

Figure 6. Simulation results

We will now consider simulation dynamics. The very first request served, at 50 ns, comes from client 1 and is only for initialization. It is followed by a PUT from client 2 and a GET from client 3, after which the buffer is empty again. Then, there are seven cycles during each of which the clients are served one after another. I.e., in each cycle two tokens are put into the buffer and only one is taken out. Consequently, shortly after 1000 ns, the 8th PUT from client 1 fills the buffer, so that client 2 cannot be served. It is skipped, and the GET request from client 3 is served instead. Hence, a place in the buffer becomes free. If client 1 were granted its request now, the schedule would be unfair because client 2 would forever continue to starve. Instead, the round-robin scheme resumes with client 2. This leads to the sequence of alternating PUT and GET requests which can be observed after the 1000 ns time stamp. Another aspect of fairness is visible on the D_OUT signal: The consumer receives tokens alternatingly from the two producers.

8. CONCLUSION

After presenting an overview of the work on synthesis from object-oriented hardware description and programming languages, we have outlined our own, language-independent approach to enable the hardware synthesis of objects. The approach is based on a static notion of object creation and interconnect. The object-oriented hardware description language Objective VHDL facilitates the practical modeling of such static object systems. Each object is synthesized into a controller / datapath circuit which provides all the services that are defined by the object's methods. Synthesis must take into account that objects may operate and interact concurrently. This has been achieved by creating arbitration and customized scheduling of concurrent service requests. All circuits have been further synthesized into gates, and have been simulated together at all levels of abstraction to show the applicability of our concepts.

9. REFERENCES

1. P. Schaumont, S. Vernalde, L. Rijnders, M. Engels, I. Bolsens. A Programming Environment for the Design of Complex High Speed ASICs. Proc. Design Automation Conference (DAC), 1998.

2. K. Bartleson. A New Standard for System-Level Design. Synopsys Inc., 1999.

3. J. L. da Silva et al. Efficient System Exploration and Synthesis of Applications with Dynamic Data Storage and Intensive Data Transfer. Proc. Design Automation Conference (DAC), 1998.

4. M. Meixner, J. Becker, Th. Hollstein, M. Glesner. Object-oriented Specification Approach for Synthesis of Hardware-/Software Systems. Proc. GI/ITG/GMM Workshop, 1999.

5. S. Maginot, W. Nebel, W. Putzke-Röming, M. Radetzki. Final Objective VHDL language definition. REQUEST Deliverable 2.1.A (public), 1997. Available on the WWW from URL http://eis.informatik.uni-oldenburg.de/research/request.html

6. R. Zippelius, K. D. Müller-Glaser. An Object-Oriented Extension of VHDL. Proc. VHDL Forum for CAD in Europe (VFE, Spring Conference), 1992.

7. W. Glunz. Extensions from VHDL to VHDL++. JESSI-AC8 report S2-SP1-T2.4-Q3, 1991.

8. B. M. Covnot, D. W. Hurst, S. Swamy. OO-VHDL: An Object-Oriented VHDL. Proc. VHDL International Users' Forum (VIUF), 1994.

84

9.　J. Benzakki, B. Djafri. Object Oriented Extensions to VHDL—The LaMI proposal. Proc. Conf. on Computer Hardware Description Languages (CHDL), 1997.

10.　P. J. Ashenden, P. A. Wilsey, D. E. Martin. SUAVE: Painless Extension for an Object-Oriented VHDL. Proc. VHDL Int'l Users' Forum (VIUF, Fall Conference), 1997.

11.　G. Schumacher. Hardware Specification and Design with an Object-Oriented Extension to VHDL. Ph.D. dissertation, Oldenburg University, 1999.

12.　S. Matsuoka, A. Yonezawa. Analysis of Inheritance Anomaly in Object-Oriented Concurrent Programming Languages. In: G. Agha, P. Wegner, A. Yonezawa (Eds.), Research Directions in Concurrent Object-Oriented Programming, MIT Press, 1993.

13.　P. Schaumont, S. Vernalde, L. Rijnders, M. Engels, I. Bolsens. A Programming Environment for the Design of Complex High Speed ASICs. Proc Design Automation Conference (DAC), 1998.

14.　T. Kuhn, W. Rosenstiel, U. Kebschull. Object Oriented Hardware Modeling and Simulation Based on Java. Proc. International Workshop on IP Based Synthesis and System Design, 1998.

15.　M. Radetzki. Synthesis of Digital Circuits from Object-Oriented Specifications. PhD dissertation, accepted by Oldenburg University, to appear in 2000.

SYSTEM LEVEL DESIGN

HW/SW Co-design

UF: ARCHITECTURE AND SEMANTICS FOR SYSTEM-LEVEL MULTIFORMALISM DESCRIPTIONS

Luis Sánchez Fernández
Dep. Tecnologías de las Comunicaciones, Universidad Carlos III de Madrid
Av. Universidad, 30, E-28911, Leganés (Madrid), Spain
luis@it.uc3m.es

Simon Pickin
Dep. de Lenguajes, Proyectos y Sistemas Informáticos,
E.U. Informática, Univ. Politécnica de Madrid
Ctra. Valencia, km 7, 28031 Madrid
spickin@eui.upm.es

Angel Groba
EUIT Telecomunicación, Univ. Politécnica de Madrid
Ctra. Valencia, km 7, 28031 Madrid
amgroba@dit.upm.es

Natividad Martínez Madrid
Forschungszentrum Informatik (FZI),
Haid-und-Neu-Stra(e 10-14, D-76131 Karlsruhe, Germany
martinez@fzi.de

Alejandro Alonso
Dep. Ingeniería de Sistemas Telemáticos,
ETSI Telecomunicación, Univ. Politécnica de Madrid,
Ciudad Universitaria, E-28040 Madrid, Spain
aalonso@dit.upm.es

J. Mermet (ed.), Electronic Chips & Systems Design Languages, 89–98.
© 2001 *Kluwer Academic Publishers.*

90

Abstract

In many industrial cases, the design of complex systems results in descriptions of the system that make use of different specification languages. In this paper we present a language, for the description of the global architecture of a system specified using such multilanguage approach.

Keywords: System architecture, multiformalism descriptions, cosimulation

Introduction

In many industrial cases, the design of complex embedded systems involves descriptions of the components of the system that make use of different specification languages. Complex embedded systems are composed of different components with different specification requirements: analog vs. discrete, software vs. hardware, electronic vs. mechanic vs. hydraulic, control dominated vs. data dominated, real time requirements, safety critical systems, etc. In contrast to those that aim at a System Level Design Language, we assert that there is no perfect language that it is suitable for any kind of system. We therefore advocate the use of the most suitable language for each component. Another typical situation in which it is possible to find multiformalism descriptions is when a system is being developed by different teams of designers with different backgrounds. Finally, another cause of multiformalism descriptions is the reuse of previously existing descriptions, which can be available in different specification languages.

When dealing with multiformalism descriptions a description of the global architecture of the system is needed: which components compose the system, how these components communicate with each other and with the global environment and what is the global semantics of the system.

In this paper we present a language (which we have called UF [Sánchez Fernández et al., 1999]) for the description of the global architecture of a system specified using several specification languages. Among the possible applications of this language we cite: documentation of the design and validation (co-simulation and consistency checking of the interfaces offered by the different components).

Also, global design (partitioning, code generation, integration in the target architecture) is possible. If no language for architectural description of the system were available, after the specification phase the design would have to be carried out individually for each of the components that form the system. Besides the possible inefficiencies of such a design flow, a disadvantage of such an approach is that it does not allow an integrated design process, and validation could be done only at the end of the design process.

We want that UF to support a wide range of specification styles. As a starting point, in [Delgado Kloos et al., 1999] we studied the features of three different specification languages: SDL [ITU-T, 1993], MATRIX$_X$ [ISI, 1996] and Statecharts [Harel, 1987]. These languages cover the description of different kind of systems: discrete (all formalisms), continuous (MATRIX$_X$), synchronous (Statecharts), asynchronous (all formalisms), control dominated (Statecharts and SDL), data dominated (MATRIX$_X$), safety critical (SDL), with hard real time requirements (Statecharts), communicating systems (SDL).

The rest of this paper is structured as follows. Section 1 presents some related work. Section 2 gives an overview of UF and the UF semantic model. Section 3 presents a simulation cycle algorithm that can be used to implement cosimulation of UF models. Section 4 shows a case study, and will also allow us to present some of the syntactic elements of UF. We finish with some conclusions to the work that has been presented.

1. STATE OF THE ART

The rapid technological advance in the field of integrated circuits has permitted the fabrication of larger and larger electronic systems. This has caused new design needs to be able to cope with the complexity of today's designs. Several techniques, methodologies and languages have appeared in recent years to ease the work of the designers.

Among the initiatives that have appeared to cope with today's design complexity, perhaps the most important is design reuse [Design & Reuse; RAPID; VSI Alliance].

Another way of dealing with complex designs is to increase the abstraction level of the specification language used. Currently, the research activities are focusing on moving from behavioural level descriptions to system-level. Several specification languages have been researched for system level design: SDL [Ismail and Jerraya, 1995], Statecharts [Harel, 1987], MATRIX$_X$ [ISI, 1996], LOTOS [Sánchez Fernández et al., 1998], Java.

Some other approaches consider the combined use of several specification languages [Kleinjohan et al., 1997; Romdhani et al., 1995]. Most of them only consider a unified global model at an intermediate level, that it is then used in the partitioning phase. This is in contrast with UF, where we provide a global model of the system at the specification level.

2. SEMANTIC MODEL

Interconnecting subsystems specified using different formalisms inevitably involves the definition of a framework in which to embed them and give unambiguous meaning to this interconnection. The main aspects of such a semantic framework are: 1) a global timing model and a definition of the relation of the timing models of each formalism to this global timing model, 2) a common data model, needed to interchange data from one formalism to another, covering at least the datatypes to be interchanged, 3) the global connectivity, that is, the specification of the communication channels between the different subsystems, and 4) global communication mechanisms, defining the structure and semantics of both the interfaces offered by the subsystems and the information exchange between them.

We have chosen a discrete global timing model even if continuous subsystems (i.e. MATRIX$_X$) are allowed in the specification. In this case, the outputs of the continuous subsystem must be sampled at the rate defined by the time scale. We have also defined an asynchronous global timing model, because we think that this choice is the more intuitive when one of the formalisms considered is asynchronous, as it is in this case (SDL).

We propose to define a common data model and map the data types of each formalism to this model. In view of the widespread acceptance which it has now achieved, and with an eye to facilitating the generation of (possibly distributed) implementations, the data part of CORBA-IDL [OMG, 1999] would constitute a good choice for a common data language.

With respect to the architecture, we had to choose between two different approaches. Either we allow the connections between the sub-components to change dynamically (a system in which each component can send a message through one of its ports to the port of any of the other components) or the connections are statically fixed. We have chosen static connections because we believe that at the level of the composition formalism the components represent the architecture of the system and because we believe this to be sufficiently flexible for the application domain.

We have also defined common communication semantics. We stipulate that all inter-subsystem communications are events, possibly with associated data exchange. For each of the formalisms a mapping to this communication semantics must be defined. For instance, in the case of MATRIX$_X$, we have defined a notion of event based on a change of value of a boolean signal.

We have taken the decision that the updating of the data outputs of each component is associated to the occurrence of events. The communication primitives in UF are events with their associated data parameters. This has to be mapped to the communication primitives in the formalisms.

The main advantage of this approach is that there are fewer communications between components than if we update data parameters whenever there is a change. We assume that data parameters are not read unless an event is produced, without loss of generality.

At the UF level we only have UF components in the specification, therefore we encapsulate the subsystems in each of the different formalisms by means of wrappers. Each basic UF component is a subsystem that comprises a wrapper encapsulating a specification described in one of the formalisms considered. They provide the information needed to define the behaviour of the UF component according to the semantic framework defined for UF in [Delgado Kloos et al., 1999]. This information is mainly concerned with the relation between the subsystem interface and the UF component interface.

To be able to build the wrappers we had to impose some restrictions to the specification languages. We need to match the interfacing scheme offered by UF with the interfacing mechanisms of the specification languages. For instance, in SDL it is possible to have dynamic creation of processes, but in UF we need static interfaces. We therefore forbid the creation of processes that communicate with the SDL environment after initialisation, again not a very serious restriction for this application domain. For more information about the restrictions imposed to the specification languages, we refer to [Sánchez Fernández et al., 1999].

With this scheme of basic components and wrappers, it is possible to translate a UF description to an intermediate format or implementation language by defining a translation of the UF primitives and the wrappers to the target notation. This scheme is possible if code generators for the specification languages supported by UF are available.

3. SIMULATION CYCLE

We assume that the global temporal behaviour of our system is asynchronous, since it is more natural to choose this less restrictive behaviour for the global system if we are to allow asynchronous components. In this case, the global system only executes a transition in moments of time when one of its components has a scheduled action and at such moments, all actions that can be scheduled in the global system are executed in zero time. In simulation, this assumes that at any time it is possible to know when their next event is scheduled.

In this section we define the semantics of UF by means of an algorithm that governs the simulation of an UF specification. The first part of this section deals with the main semantic aspects following from this assumption. The second part of this section deals with the implementation of a simulation cycle, explaining how asynchronous, synchronous and continuous time simulators participate in this cycle.

For simplification, we assume here that inter-subsystem parallelism is implemented through interleaving so that if, at any given moment, more than one subsystem has scheduled actions, these actions are interleaved. There are two main choices with respect to the granularity of this inter-subsystem interleaving:

- fine-grain interleaving: interleaving of individual transitions from each subsystem

- coarse-grain interleaving: interleaving of maximal sequences of (zero-time) transitions from each subsystem (c.f. the Statemate notion of "superstep")

That is, whether, during its turn, a subsystem is only allowed to execute a single transition or it is allowed to execute as many transitions as it can execute in that time instant. We assume a zero delay model for inter-subsystem communications. Note that the output from one subsystem can be used as input in another subsystem in the same time instant, so that the problem of zero-time loops involving more than one subsystem may arise.

A subsystem executing according to a synchronous time scheme (Statecharts) can be embedded in an asynchronous global time scheme by ensuring that it is only scheduled once at each instant for which it has executable actions, using the outputs of the previous instant as its inputs in the current instant.

A continuous time subsystem ($MATRIX_X$) can be embedded in an asynchronous time scheme with zero-delay communications and where all actions of the discrete-time formalisms take zero-time by considering that:

- the values of its inputs from the rest of the system are constant between time instants where actions are scheduled in some part of the global system, and change "after" the execution of all the zero-time actions of the discrete time formalisms scheduled for that instant

- the values of its outputs to the rest of the system are constant between time instants where actions are scheduled in some part of the global system, and change "before" the execution of all the zero-time actions of the discrete time formalisms scheduled for that instant

- the continuous subsystem can schedule outputs to the rest of the system at future discrete global instants thus influencing the global time advance. This point needs a notion of communication event to be defined for the continuous subsystem.

3.1 IMPLEMENTATION OF A SIMULATION CYCLE

The simulation cycle assumes that all the components in each of the formalisms communicate when they have scheduled their next events. In some cases, this information is not available from the simulator. In these cases, to obtain this information it may be necessary to simulate until an event appears, take note of the simulation time, and then rollback to the initial simulation point.

For a cosimulation with asynchronous temporal behaviour, if more than one simulator proposes an action at any one instant, the backplane must decide on how to schedule these actions (implementing parallelism as interleaving), assuming the granularity of interleaving has already been defined. The cosimulation backplane control loop is as follows:

1. advance time in each of the simulator instances (executing the necessary actions in the continuous and synchronous simulator instances) to the earliest of the following:

    - the earliest global time at which one of the asynchronous instances has a scheduled action

    - the earliest global time at which one of the synchronous or continuous instances has an external communication to make.

2. schedule a step in one of the simulator instances (recall the notion of granularity of interleaving), in the case where more than one simulator instance proposes an action at this time; if this is a synchronous instance, mark it as already having evolved in this instant so that it will not be evolved again without evolving time

3. if this step involves sending to an asynchronous simulator instance, deliver the communication and interrogate this receiver instance to see if it now has scheduled actions in the current instant; if this step involves sending to a continuous simulator instance ensure that the communication is not available to be consumed by the receiver instance until time advances

4. continue scheduling steps until none of the simulator instances can advance without evolving time

We are currently working in a cosimulation tool that will allow us to simulate UF descriptions. For this task we are developing our own cosimulation backplane, based on CORBA [OMG, 1999].

4. CASE STUDY

In this section we present a case study that will allow us to show how a UF description looks like and the main syntactic elements available in UF. We do not go into all the details of the case study, but only show the global architecture of the system. The complete description of the case study can be found in http://www.it.uc3m.es/~comity/demo/demo.html

Our system is a controller for opening and closing car windows. All the four windows in the car are controlled. The people on the front of the car can control all the car windows. The people on the rear of the car only control their own windows. In case of conflict, the driver's commands have priority. We decided to model a simple system in which every window is controlled by a processor embedded in its corresponding door, and a CAN bus is used for the communication between the different processors of the system. The model

Figure 1.1 architecture of the window lift

Figure 1.1 architecture of the window lift

of the system also includes the window motors. This allows us to model the actions/reactions between the motor and the controller.

We decided to use SDL, MATRIX$_X$ and Statecharts to develop this model. SDL is a language suitable for specification of communication systems, so we decided to use SDL for the CAN bus. MATRIX$_X$ is adequate for continuous systems, and therefore we used it for the motor. Finally, control parts are very well described in Statecharts. Therefore, we used Statecharts to model the processors that control the window lift system. We can see in figure 1 the global architecture of the system.

The model is composed of four instances of a MATRIX$_X$ model for the motor, four instances of the Statecharts model of the controller and one SDL model for the CAN bus. Below we can see the UF global description of the system.

```
UF_specification Window_lift;
UF_component CAN_bus
Interface
fr_ind, fl_ind, rr_ind, rl_ind:  out (msg_type:  long);
fr_req, fl_req:  in (msg_type:  long);
use CAN_bus;
UF_component Controller
...
UF_component Motor_4
...
```

First we list the components that compose the description and the interface of each of the components. Then we link each component with its UF behaviour description that can be in the same file or in a library (use statement).

```
event fr_ind, fl_ind, rr_ind, rl_ind:  (msg_type:  long);
event fr_req, fl_req:  (msg_type:  long);
```

We also have to define the events (and the attached data) exchanged by the components in the model (event statement).

```
Behaviour
Motor_4 ( ... );
CAN_bus(fr_ind, fl_ind, rr_ind, rl_ind, fr_req, fl_req);
Controller (m2_Blocked,m1_Blocked,m4_Blocked,m3_Blocked,
fl_ind, fr_ind, rl_ind, rr_ind, ...);
```

Finally we instantiate the components and define the connections between the models. This is done by assigning the same event name to the corresponding input and output of two components. For instance, fr_ind, fl_ind, rr_ind, rl_ind are four outputs of the CAN_bus UF model and therefore, should be four inputs of the Controller UF model.

We have to define a UF component that encapsulates each of the subsystems defined in the three notations. As an example below we present the UF component for the CAN bus. The UF specification defines a mapping between the UF interface and the SDL interface and also the SDL file where the SDL description of the CAN bus is stored.

```
UF_specification CAN_bus
Interface
fr_ind, fl_ind, rr_ind, rl_ind:  out (msg_type:  long);
fr_req, fl_req:  in (msg_type:  long);
Behaviour (SDL)
URL: http://www.it.uc3m.es/~comity/can\_bus.pr
Input signals:
req:  fr_req => SDL_process(CAN_controller,1),
      fl_req => SDL_process(CAN_controller,2);
Output signals:
ind (process CAN_controller,1):  fr_ind;
...
```

5. CONCLUSIONS

In this paper we have presented a language for describing the global architecture of a multiformalism description. Our architectural language provides a semantic framework where a wide spectrum of formalisms can be integrated including continuous and discrete (with synchronous and asynchronous time models) notations.

Currently we support three specification languages inside UF: MATRIX$_X$, SDL, Statecharts. However, our notation is intended to be of general application, so many other specification languages could also be integrated. A case study has been used to show our approach. This case study is based on an industrial multiformalism case study that has been developed by BMW.

Acknowledgments

This research has been partially carried out in ESPRIT project No.23015 COMITY (Co-design Method and Integrated Tools for Advanced Embedded Systems), and in the "Accion Integrada hispano-alemana" HA97-0019 "Metodología de diseño de sistemas reactivos", a collaboration with Tec. Univ. of Munich, Univ. of Augsburg and Med. Univ. of Lübeck, funded by the Spanish Government. The authors wish to acknowledge the collaboration of the COMITY consortium partners: Aerospatiale, ISI, TIMA, Verilog (France), BMW, C-Lab (Germany) and Intracom (Greece).

References

Delgado Kloos, C., Pickin, S., Sánchez, L., and Groba, A. (1999). High-Level Specification Languages for Embedded System Design. In Jerraya, A. and Mermet, J., editors, *System Level Synthesis*, pages 137–174. Kluwer Academic Publishers.

Design & Reuse. http://www.design-reuse.com.

Harel, D. (1987). Statecharts: A visual formalism for complex systems. *Science of Computer Programming*, 8:231–274.

ISI (1996). System Build Reference Guide. Integrated Systems Incorporated.

Ismail, T. and Jerraya, A. (1995). Synthesis steps and design models for co-design. *IEEE Computer*, 28(2):44–52.

ITU-T (1993). Recommendation Z.100. CCITT Specification and description language SDL.

Kleinjohan, B., Tacken, J., and Tahedl, C. (1997). Towards a complete design method for embedded systems using Predicate/Transition-Nets. In Delgado Kloos, C. and Cerny, E., editors, *IFIP Conf. Hardware Description Languages (CHDL)*, pages 4–23. Chapman & Hall.

OMG (1999). CORBA/IIOP 2.3.1 Specification. The Object Management Group. http://www.omg.org/corba/corbaiiop.html.

RAPID. Reusable Application-Specific Intellectual Property Developers. http://www.rapid.org.

Romdhani, M., Hautbois, R., Jeffroy, A., de Chazelles, P., and Jerraya, A. (1995). Evaluation and composition of specification languages, an industrial point of view. In *IFIP Conf. Hardware Description Languages (CHDL)*, pages 519–523.

Sánchez Fernández, L. et al. (1998). Hardware-Software Prototyping from LOTOS. *Design Automation for Embedded Systems*, 3(2/3):117–148.

Sánchez Fernández, L. et al. (1999). Definition of the Composition Formalism. Deliverable D22.2 of the EP 23015 COMITY Project.

VSI Alliance. http://www.vsi.org.

Automatic Interface Generation among VHDL Processes in Hardware/Software Co-Design

Cristiano C. de Araújo and Edna Barros
Departamento de Informática, UFPE, Recife, Brazil

Key words: Interface generation, hardware/software co-design, system synthesis

Abstract: The modern electronic systems are becoming more complex, requiring shortest
design time and demanding a decreasing in design cost. To achieve this,
techniques for supporting hardware/software co-design have been developed
in order to permit the joint specification, design and synthesis of mixed
hardware/software systems. But the use of co-design brings to the designer a
very hard and error prone task, the implementation of the interface between the
hardware and software parts. This way the designer has to deal with three
projects : the hardware, the software and the interface. The main goal of this
work is the development of a methodology for automatic interface generation
between concurrent processes, releasing the designer of this task and letting
him or her free for spending efforts on the design of the system itself. This
work   focuses on the automatic interface generation between concurrent
VHDL processes communicating through message passing. The proposed
approach includes the development of a library of generic communication
components, the definition of VHDL constructors for synchronous
communication, as well as techniques for automatic VHDL code generation
with synchronous communication, which can be simulated and synthesised by
conventional CAD tools.

1.   INTRODUCTION

With the growing complexity of the digital systems and the need for reducing the time to market, techniques for supporting hardware/software

99

J. Mermet (ed.), Electronic Chips & Systems Design Languages, 99–110.
© 2001 *Kluwer Academic Publishers.*

100

co-design have been [1]developed in order to permit the joint specification, design and synthesis of mixed hardware/software systems [5,6]. Such systems consists of common-off-the-shelf and ASIC components and have a variety of implementation technologies and interfaces, and a wide range of real-time data rates. The need for early prototypes to validate the specification and to provide the customer with feedback during the design process is another key factor motivating hardware/software co-design.

Some tools and methodologies supporting hardware/software co-design have been developed in the last years [5,6,7,8,9,10,11,12]. In most of them, however, once the initial description was partitioned, the interface between the hardware and the software components is synthesised by hand or in a semi-automated way.

This work takes into account the PISH co-design system, which allows the partitioning of occam descriptions by considering hardware/software trade-off but also distinct hardware implementations [11]. Additionally, the correctness of the partitioning process can be assured through the use of formal verification techniques, in a constructive way [12]. The partitioning output is then a set of communicating modules, some of them to be implemented in hardware and others to be implemented in software. This set of modules represents a first system prototype, that is a virtual prototype, and the next step is the mapping of this virtual prototype into a real one with the synthesis of hardware and software modules as well as the interface between them. This step also called co-synthesis is a very time consuming and error prone activity, and in the PISH co-design system it has been done by hand.

The complexity of the interface generation depends on the flexibility of the target architecture. The most systems with automatic partitioning taken into account a pre-defined target architecture, which makes the interface generation easier. But also in this case, automatic interface generation can be not easy due to the semantic gap between the specification mechanisms representing the virtual and the real prototypes, respectively. Due to this fact, techniques for automatic interface generation is a feature of a small number of co-design systems [2,16].

An important support for interface generation is the automatic communication generation during the partitioning process. When the communication among modules is made explicit and assured to be correct, the mapping of the virtual prototype into real one can be done in a most natural and correct manner. The main goal of this work is the development of techniques for automatic interface generation in the PISH co-design system. The interface generation system must generate the interface between hardware and software, the interface between hardware modules and the

[1] This work has been supported by CNPq (Project nr. 52 1869/96.0)

interface between software modules when they are running on distinct components (microprocessors). This work aims to propose a technique for interface generation between hardware modules, which are described in VHDL.

This paper is organised as follows: the next section gives the related works, then an overview of the PISH co-design system including the approach for automatic interface generation is presented. A technique for interface generation between hardware modules is described in section 4. How this technique has been implemented is discussed in section 5. Section 6 illustrates an example of interface generation. Some conclusions and future works are presented in section 7.

2. RELATED WORKS

The interface in POLIS approach [17] implements a domain specific communication mechanism between a set of co-design finite state machines (CFSM's). This mechanism uses asynchronous communication and is based on event detection. The main problem with this approach is that it uses different approaches and protocol for the three different interfaces : hardware/hardware, hardware/software and software/software.

In [18] interfaces for synchronous dataflow (SDF) are automatically generated. This methodology uses a hierarchical approach to interface generation. A layered representation of the interface is given where in the first layer the communication is represented by abstract unidirectional links connecting source and sink nodes. These links are virtual channels. A second layer, implementation layer, is composed of computing elements (processors, microcontrollers, FPGA's), buses and memories. The automatic interface generation is performed by mapping the objects of the abstract layer to the implementation layer using the HASIS tool. This mapping is done by code generation not using component libraries.

In [3] an intermediate abstract architecture is also used. This architecture is composed of processing elements and point-to-point unidirectional channels. The processing elements can be hardware components or processors. In the latter case a hardware *wrapper* that encapsulates the software component as a hardware one is added to the processor core. This hardware *wrapper* implements the communication interface to its eternal environment. Unlike the previous one the automatic interface generation is done by choosing the right hardware *wrapper* components stored in libraries.

Our approach uses a domain specific mechanism like POLIS, but unlike it, synchronous communication is supported as [3]. Another key aspect is that an extended VHDL set is used, this superset includes synchronous communication statements.

3. TOWARDS AUTOMATIC INTERFACE GENERATION IN THE PISH CO-DESIGN SYSTEM

The PISH co-design system uses occam as specification mechanism and comprises all phases of the design process including partitioning, interface synthesis, hardware synthesis and software compilation as depicted in Figure 1. The occam specification is partitioned into a set of processes. A pre-defined architecture is taken into account, which includes only one software component. In this way, one process is going to be implemented in software and the others, in hardware.

The partitioning is done in such way that it can be formally verified that the partitioned description has the same semantics as the original one. Processes for communication purposes are also generated during the partitioning phase [13]. As mentioned, this is an important support for interface generation.

After partitioning, the processes to be implemented in hardware are synthesised and the software processes compiled, as well as the interface hardware/software and hardware/hardware. In the current version, this interface is generated by hand. An approach for automatic generation is the main goal of this work and techniques for interface generation between hardware modules will be focused in this paper.

3.1 Automatic Interface Generation in the PISH Co-design System

As mentioned, the partitioning phase is closely related to the automatic interface generation. Depending on the specification mechanism used to represent the virtual prototype and on the target architecture taken into account, the interface generation can be more or less complex.

In the PISH co-design system occam [14] is used as description language. The main reason to use occam is that, being based on CSP [15], occam has a simple and a elegant semantics, given in terms of algebraic laws.

In order to preserve the semantics of the original description, the partitioning process is done by applying a set of transformation rules in an occam description, guided by the results of a cost analysis based on clustering techniques [12,13].

The output of the partitioning phase is a set of concurrent communicating processes, which communicates through processes generated only for this purpose. The communication among processes is generated by using formal methods. It can be assured that the semantic of the initial description has been preserved and the processes communicate in a correct way (no deadlock was introduced). This feature is a important support for the interface generation in the next step, since the communication among processes has been made explicit and is correct. An approach for interface generation is being developed, which can be seen in Figure 1.

The interface generation between hardware and software includes the generation of device drivers, which allows the link between processes to be implemented in software, and processes in hardware.

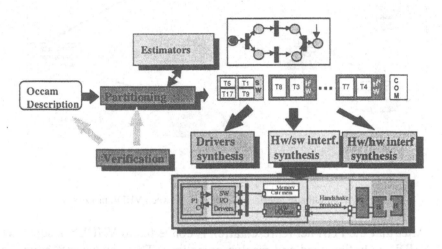

Figure 1. The PISH co-design system including interface generation

The generation of device-drivers is done for components in a library and takes into account the corresponding communication protocol. In the hardware side, additional circuits must be generated in order to make it possible the hardware see the software component as another hardware component. The link between the software component and the hardware is done by the communication unit. The interface between hardware components is done by using a pre-defined protocol and the communication is synchronous through channels. This feature allows an implementation, which is compatible with the occam semantics.

4.   AUTOMATIC INTERFACE GENERATION IN VHDL DESCRIPTIONS

The design flow for automatic generation of VHDL code from occam processes to be implemented in hardware with explicit communication can be seen in Figure 2. First we have occam specifications describing the hardware processes and their communications. These specifications are inputs for the occam to Petri Net converter. Petri Net is used as intermediate format in the PISH co-design system [1].

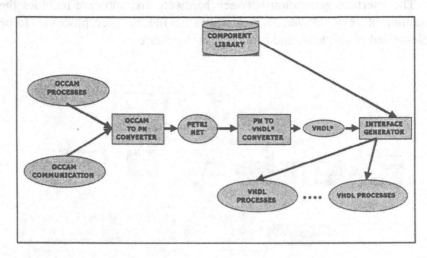

Figure 2. Automatic Interface Generation between VHDL processes

After that the Petri net representation is converted to VHDL*, a super set of VHDL including send and receive operations. These statements have the CSP communication semantics where both sender and receiver must synchronise before the data transfer takes place. In the VHDL* description the hardware processes communicates by sending and receiving messages. A VHDL description of such processes is generated by the interface generator. The resulting description can be synthesised by available synthesis tools.

In the used communication model , each process communicates with another one through unidirectional channels, bi-directional communication can be implemented by using two channels. A channel is composed of two components, a send and a receive one, which implements the communication protocol. The components and communication protocol are based on the work developed at IMEC, Belgium [2,3].

4.1   Model of the Hardware Processes in VHDL*

In order to make transparent the communications details for the designer, we have extended the VHDL language with two new sequential statements. A send statement and a receive statement. The syntax of the send statement is *channel ! expression* and of the receive statement is *channel ? variable*. The semantics of these statements is the CSP semantic explained previously.

The main idea of our approach is to model the hardware processes as FSM's (finite state machines) since the most of today's digital hardware operates synchronously under a clock signal control [5]. The FSM model of a hardware process in VHDL* can be seen in Figure 3. The description includes a type declaration, where the states are listed and a process statement where the outputs and next state are updated at each clock transition. In each state transition a series of sequential statements can be performed including the communication statements (send/receive).

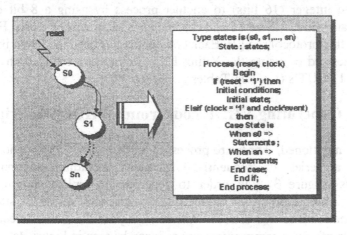

Figure 3. FSM Model of a Hardware Process

5.   IMPLEMENTATION OF SYNCHRONOUS COMMUNICATION

The implementation of synchronous communication is done by using parameterised components of a library, including the sender and the receiver. As mentioned, they implementing the protocol of communication, so the process don't have to care about communications details. By taking these components into account, the interface generator modifies the process

106

description in VHDL* in order to implement the send and receive operations. These modifications must guarantee that the CSP semantics is preserved.

5.1   Communication Components

The communication components were implemented by using the model and protocol proposed by [2]. According to this protocol, the components at both sides must be activated in order to start the communication. When the communication is finished both components must set a corresponding signal. In order to cope with distinct data types, the components have two parameters: process data width and channel data width. The process data width is the number of bits of the data to be sent or received. The channel data width means the number of bits that will be transmitted at a time. This parameter must be given by the designer. For example, if one process has to transfer an integer (16 bits) to another process by using a 8 bit channel, process data width will be 16 and channel data width will be 8 bits. For more details of this protocol see [2]. Each component has been modelled in VHDL and synthesised by using the Xilinx F1.4 Foundation tools. Both of them occupied 11 LUT's (Lookup Tables).

5.2   Generating VHDL code from a VHDL* description

As mentioned, hardware process is modelled as a FSM, whose states contains a series of sequential statements including communication statements (Figure 3). In order to generate an implementation for the communication action, we divided the statements inside a state with communication into two classes: *statements before the communication* and *statements after the communication,* as it can be seen in Figure 4a.

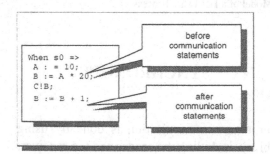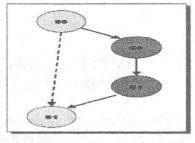

Figure 4 – (a) The state of FSM and (b) the additional states

The strategy for generating VHDL code with communication consists in replacing each communication statement by a set of state assignments in order to activate the signals of the communication components in a correct way. Two new states for each send/receive statement must be insert in the process. The first state activates the communication component and sets a signal when communication has finished. The second state includes all sequential statements after the communication. In Figure 4a we have the original sequence of states s0, s1 where s0 contains one communication statement. So the two states c0 and c1 have been insert, c0 activates the communication component and c1 contains the "after communication" statements of the original s0 state. The modification of the VHDL* code for including new states as well as the communication components can be done automatically by using a VHDL parser implemented in Java [4, 17].

6.   AN EXAMPLE

In order to make clear the proposed approach, the whole process is described in this section starting from a set of occam processes to be implemented in hardware until the generation of a VHDL description. As depicted in Figure 5a the example, is composed of two concurrent processes.

The first process, P0, compares the signal A with the internal variable *temp_a* , which contains its previous value. If they are different, it sends the new value of A through the channel "C". The second process, P1, receives the value from the channel "C", stores it on a variable *x,* and does a simple AND operation resulting on the signal Y. In Figure 5b we have the occam specification of the two processes as well as their Petri net representation. The occam constructor "PAR" defines the two concurrent processes, each one composed of sequential primitive processes such as assignments and input/output. In the Petri net description the two processes are represented by places and transitions [1], the places labelled with P0 belongs to the process P0, the ones marked with P1 to process P1. One special transition labelled with "T" represents the communication between the two processes [1]. In the Figure 5c the VHDL* description of the process P0 is depicted. This description includes send and/or receive constructors for communication. Standard VHDL code is generated by inserting some new states implementing the communication as well as the communication components. Figure 5d shows the VHDL description with the new states for communication. Figure 5e shows the declaration and instantiation of the sender component.

As a result of the conversion from VHDL* to VHDL, the original FSM in VHDL* is increased with two new states for each communication statement, and communication components (sender and receiver) for each channel is included in the code. The parameterised components have a VHDL code of about one hundred lines and occupy 11 LUT's in a Xilinx XC4003 FPGA for a 16 bit communication channel.

108

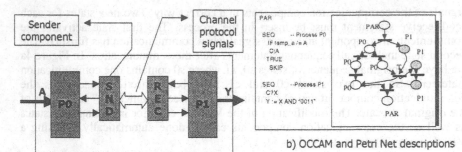

a) structural view of the example

b) OCCAM and Petri Net descriptions

```
Entity P0 is
port (
  rst, clk, init_process, A : in std_logic;
  end_process : out std_logic);
end P0;

architecture P0_arc of P0 is
  type states is (init, end, s0, s1, s2, s3);
  signal state : states;
begin
  P0_process : process (rst, clk)
    variable temp_a : std_logic;
  begin
    if (rst = '1') then state <= init; end_process <= '0';
      temp_a := '0';
    elsif (clk = '1' and clk'event) then
      case state is
        when init =>
          if (init_process = '1') then
            state <= s0;
            end_process <= '0';
          end if;
        when s0 =>
          if (temp_a /= A) then state <= s3;
          else state <= s1;
          end if;
        when s1 => state <= s2;
        when s2 => state <= end;
        when s3 => C ! Temp_a; state <= s2;
        when end => end_process <= '1'; state <= init;
      end case;
    end if;
  end process;
end P0_arc;
```

c) VHDL*

```
when s3 => state <= C_com_0;
when C_com_0 =>
  if (not C_activated) then
    if (C_ready = '1') then
      C_activated := false;
      C_com_started := false;
      state <= s3_0;
    end if;
  else
    if (C_ready = '0') then
      C_com_started := true; end if;
  end if;
when s3_0 => state <= s2;
```

d) states added

```
u_C : sender
generic map (MAX_BITS_IN => 4;
            MAX_BITS_CHANNEL => 4)
port map (
  reset    => rst,
  send     => C_send,
  clk      => clk,
  rec_rdy  => C_rec_rdy,
  send_rdy => C_send_rdy,
  data_prc => C_ data_prcm,
  data_ch  => C_ data_ch,
  ready    => C_ready
  );
```

e) sender component added

Figure 5. Descriptions in OCCAM, Petri Net, VHDL.* and VHDL

7. CONCLUSIONS

In this paper we have presented a technique for automatic interface generation between VHDL processes. This technique is part of a more general approach for automatic interface generation in the PISH co-design system. In the PISH system the resulting partitioned system is a occam description including processes to be implemented in hardware, processes to be implemented in software and processes for communication purpose. In this paper we have focused only the interface generation between hardware processes. The communication among such processes is done by message passing, which could be represented by adding two new constructors in VHDL, *send* and *receive*. The generation of VHDL* code is done by using a based on Petri net approach and the generated set of VHDL* processes are modeled as FSM. During the interface generation VHDL code with communication is generated by using components of a library. The code generation strategy includes new states in the VHDL* description, as well as signals and components declarations for implementing synchronous communication. The communication components can be customized for the application by setting some parameters, but, at the moment, they implement a pre-defined communication protocol. A strategy for taking into account more generic communication components is a future work. An approach for drivers generation at the software side, as well as the communication units for the hardware/software interface is under development.

REFERENCES

[1] P. R. Maciel and E. Barros, *Capturing Time Constraints by Using Petri-nets in the Context of Hardware/Software Codesign*, Proceedings of the 7th IEEE INTERNATIONAL WORKSHOP ON RAPID SYSTEMS PROTOTYPING, IEEE-Press, 1996.

[2] B. Lin, S. Vercauteren, H. De Man,`` Constructing Application-Specific Heterogeneous Embedded Architectures for Custom HW/SW Applications", ACM/IEEE Design Automation Conference, June 1996.

[3] B. Lin, S. Vercauteren, H. De Man, `` Embedded Architecture Co-Synthesis and System Integration", International Workshop on Hardware/Software Codesign, March 1996.

[4] A. A. Sarmento, J.H.C. Fernandes and E. Barros, *HardWWWired: Using the Web as Repository of VHDL Components,* submitted to the FDL'99.

[5] D. Gajski and F. Vahid, *Specification and Design of Embedded Hardware-Software Systems*–IEEE Design and Test of Computers, pp.53-67, Spring 1995

[6] T. BenIsmail, M. Abid, K. O'Brien and A. Jerraya,,*An Approach for Hardware/Software Codesign*– Proceedings of the RSP 94– Grenoble, França, 1994 - RSP94

[7] A. Kalavade , E. Lee, *A Hardware-Software Codesign Methodology for DSP Applications* – IEEE Design and Test of Computers, pp. 16-28, September 1993

[8]D. E. Thomas, J. K. Adams, H. Schmit, *A Model and Methodology for Hardware/Software Codesign*– IEEE Design and Test of Computers, pp. 6-15, September 1993

[9]R.K. Gupta. , C.N. Coelho , G. De Micheli, *Synthesis and Simulation of Digital Systems Containing Interacting Hardware and Software Components*– Proceedings of the 29th Design Automation Conference, 1992

[10]R. Ernst , J. Henkel, T. Benner, *Hardware-Software Co-Synthesis for Microcontrollers*– IEEE Design and Test of Computers, pp. 64-75, December 1993

[11]E.Barros, *Hardware/Software Partitioning using UNITY* PhD thesis, Universitaet of Tuebingen, 1993.

[12]E. Barros and A. Sampaio,. *Towards Probably Correct Hardware/ Software Partitioning Using Occam.* In Proceedings of the Third International Workshop on Hardware/Software Codesign, (1994) 210-217, IEEE Press.

[13]L. Silva, A. Sampaio and E. Barros, *A Normal Form Reduction Strategy for Hardware/Software Partitioning. In the Proceedings of the Conference Formal Methods Europe'97*

[14]D. Pountain and D. May, *A Tutorial Introduction to OCCAM Programming.* Inmos BSP Professional Books, (1987).

[15]C. A. R. Hoare, *Communicating Sequential Processes* Prentice-Hall, 1985

[16]P. Chou, R.B. Ortega and G. Borriello, *The Chinook Hardware/Software Co-synthesis System.* Proceedings of the 8th International Symposium on System Synthesis. 1995.

[17]F. Balarin, A. Jurecska, and H. Hsieh et al. *Hardware-Software Co-Design of Embedded Systems : The Polis Approach.* Kluwer Academic Press, Boston, 1997.

[18] M. Eisenring and J. Teich, *Domain-Specific Interface Generation from Dataflow Specifications,* Proceedings of the 6th International Workshop on Hardware/Software Codesign, March 1998.

System-Level Specification and Architecture Exploration : an Avionics Codesign Application

François CLOUTE[1], Jean-Noël CONTENSOU[1], Daniel ESTEVE[1], Pascal PAMPAGNIN[2], Philippe PONS[2], Yves FAVARD[2]

[1]*Laboratoire d'Electronique LEN7 de l'ENSEEIHT-2 rue Charles Camichel 31071 Toulouse Cedex 7, France*

[2]*AEROSPATIALE MATRA Airbus, Direction Systèmes et Services,316, route de Bayonne, 31060 Toulouse Cedex 3, France*

{cloute, contenso, esteve} @len7.enseeiht.fr

{pascal.pampagnin, philippe.pons, yves.favard} @avions.aerospatiale.fr

ABSTRACT: Digital designers have been used to mixing programmable and specific hardware components for algorithms implementation. However, with the growing complexity of systems, a computer-aided co-design methodology becomes essential. This methodology relies on an executable system-level specification which abstracts the implementation level and enables to perform architecture trade-offs, before the automatic synthesis step. This paper presents an application of the avionics domain: an interface system of the standard ARINC communication protocol. The codesign approach is based on the POLIS framework. A distributed system is specified with the synchronous language Esterel, combined with an asynchronous communication model . The target architecture mix hardware/software components commonly used for the implementation of embedded controllers.

1. INTRODUCTION

The major constantly growing factor that limits the development of complex systems is not the silicon technology manufacture, but the lack of a system-level design methodology. The increasing widespread of embedded systems in the domains of vehicle, avionics, and communication emphasizes that need.

Unlike a general-purpose computer, an embedded system has to realize a well-defined set of specific tasks. The required specialization should alter minimally its flexibility, to get a maximum design reuse [6].

111

J. Mermet (ed.), Electronic Chips & Systems Design Languages, 111–120.

Thus an embedded system typically consists of some VLSI hardware components, like ASIC or FPGA, and software supported by standard programmable components, like RISC or DSP.

In a modern commercial aircraft, the avionics (*i.e.* the set of on-board hardware and software electronics equipment) consists of about a hundred of computers communicating between them and the environment. Each of these computers is dedicated to a specific avionics function. Such critical systems require a certified development.

Currently, the design of an embedded system is not optimal:

1. the system specification is written in a natural language, eventually without abstraction of architectural details;
2. the architectural decisions are made a priori, following the architect's experience or/and the past product versions;
3. the hardware and software parts are developed too separately;
4. the software is tested only after the hardware/software integration on a real prototype;

An hardware/software codesign methodology aims at solving all those issues [5], [7]. Codesign is defined as a methodology for designing software and hardware concurrently, thus reducing the design time and time-to-market. Hardware/software codesign of embedded systems includes co-specification, hardware/software partitioning, architecture selection, co-synthesis and co-verification.

There are different approaches of codesign, related to the type of the target applications. The taxonomy of embedded systems distinguishes two main domains: control-oriented and data-dominated applications.

In data-flow applications, *e.g.* digital signal processing, the behavior of the system is scheduled at a fixed rate, and the main complexity of the design comes from the mathematical operations on data. In control-oriented reactive applications, the system reacts continuously to the environment. Then, the monitoring of the different tasks is crucial, especially as there are real-time constraints [2], [11].The distinction is trivial, since very complex systems deal with both. But designing them requires a separate point of view.

This paper describes the hardware/software codesign of an avionics embedded control-dominated system: the ARINC protocol interface system. The next section provides some background about the ARINC protocol interface. Section 3 considers the system specification with an Esterel overview. Section 4 presents the POLIS codesign approach. Section 5 highlights some experimental results about the hardware/software design space exploration. Section 6 concludes and discusses future work.

2.   THE AVIONICS ARINC PROTOCOL INTERFACE

The ARINC (Aeronautical Radio Inc.) is an international standard which specifies the communication protocol between the different embedded systems on board. Thus embedded systems designed by different manufacturers can communicate each other in the same aircraft. The standard protocol defines the type of data frames and the exchange format of those data. However, the requirements do not force the implementation.

The ARINC protocol is a serial communication protocol with a rate of 100 Kbits per second. Data packets are 32 bits, added to 4 bits for the synchronization. A data packet consists of an 8-bit identification field, a 23-bit data field, and one parity bit. An ARINC bus is a set of channels, each carrying data packets.

The ARINC interface system is in charge of the acquisition of the data packets received on several parallel input channels. For each channel, after the synchronization bits, the recognition and the parity checking, the message is received. For each message true to the ARINC pattern, an address is computed from the identification field to store the data field and dating information. A pre-programmed memory is used for the addressing. Concurrently, the environment asynchronously requests the ARINC interface to return the available data.

The ARINC interface system represents a critical real-time embedded system, both with a complex control based on data values and soft/hard timing constraints.

Codesign methodology aims at providing a rigorous design, ended in a final prototype with adequate hardware/software architecture. The methodology we used is supported by POLIS [1], a codesign framework developed at the Berkeley University. The specification language is Esterel [3], a reactive synchronous programming language from the INRIA Institute of Sophia-Antipolis, France.

3.   THE ESTEREL SPECIFICATION

Esterel is a textual, imperative, synchronous language, oriented towards the specification of control-dominated reactive systems.

Programming in Esterel is facilitated by a concurrent modular decomposition, and an explicit definition of the control by the use of program constructs for concurrency, preemption, exception.

114

Unlike other synchronous languages [8], like Lustre [9] or Signal [10] dedicated to the computational systems, Esterel is restricted to the addition of a header file in C for the definition of procedures or functions.

In Esterel, the basic element is the signal, valued or not, emitted by the system or the environment, and at the same time received, according to the synchronous semantics. Time is a multiform concept, based upon the nature of the signals (time, distance, temperature, etc).

The functionality of the ARINC interface was decomposed into 10 different modules, with some eventually instantiated more than once. The first step was to write each module in Esterel and to verify it by simulation with a graphical debugger.

The figure 1 presents the functional decomposition of our model of the ARINC interface system.

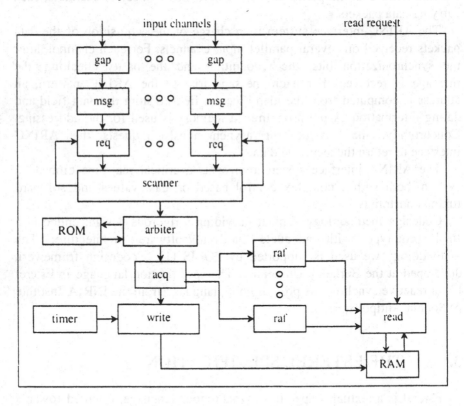

Figure 1. Model of the system

The data packets coming from each channel are detected and tested with respect to the ARINC pattern by each concurrent modules GAP, MSG and REQ. Any accepted packet awakes the process of the module SCANNER, which under some conditions, enables the access to the pre-programmed memory with a priority mechanism fixed by the module ARBITRER. The module ACQ performs the storage addressing, annotated with both dating coming from DATE, and other information which commands the modules RAF, and WRITE to make the relevant processing to the data in RAM. The module READ reacts also to any read request from the environment.

For example, the code shown in figure 2 gives a part of an Esterel module. The reactive behavior is an infinite loop that tests the presence of the signal START_RAF. Inside the loop, a down counting is performed from the last value of the signal START_RAF, with an exception handling if the internal variable VALEUR is equal to zero. The corresponding trap maintains a signal BIT_DEFAUT_DE_RAF emitted, unless a new signal START_RAF is received.

```
 . . .
   %loop
   every START_RAF do
     var VALEUR : integer in
       %initialization
       VALEUR := ?START_RAF;
       %exception handling
       trap DEFAUT_DE_RAF in
         every CLK_20MS do
           VALEUR:=VALEUR-1;
           if VALEUR=0 then exit DEFAUT_DE_RAF; end if
         end every
       handle DEFAUT_DE_RAF do
         sustain BIT_DEFAUT_DE_RAF;
       end trap
     end var
   end every
 end module
```

Figure 2. Esterel code

We can compile an Esterel program into a deterministic and sequential finite-state-machine. The table 1 shows the complexity of the ARINC interface, by analogy with the finite-state-machine parameters.

A top-level description was specified in Esterel, with the acquisition of four input channels, and a simple model of pre-programmed memory which infers two storage addresses. The Esterel synchronous specification was verified by simulation. As the final ARINC interface system has to cope with up to 56 input channels concurrently, hardware should be essential to meet the timing constraints. Thus our Esterel modules were entered in the POLIS codesign flow.

CFSM	instances	states	functions	signals	variables	actions	halts	calls
Gap	4	4	0	5	4	7	3	23
Msg	4	4	6	10	16	3	3	76
Req	4	4	0	3	2	3	3	10
Scanner	1	11	0	8	4	8	10	23
Arbiter	1	3	7	26	46	69	2	546
Acq	1	3	0	11	12	14	2	28
Date	1	3	0	2	3	7	2	14
Raf	2	5	0	4	4	7	4	24
Write	1	3	1	6	10	7	2	11
Read	1	10	0	9	10	12	6	537

Table 1: Functional decomposition of the ARINC interface with a hierarchical concurrent

4.   THE POLIS CODESIGN ENVIRONMENT

The POLIS codesign framework is oriented towards control-dominated reactive embedded systems, with generic target architecture composed of one microcontroller and some hardware components.

The POLIS environment is based on a formal model called Codesign Finite State Machine (CFSM). A system is represented by a network of CFSMs, with an asynchronous communication model between CFSMs. An Esterel module, locally synchronous specifies each CFSM.

The POLIS design flow is illustrated by the figure 3, and the main steps are described below:

1. Translation of the system-level language like Esterel into the CFSM model;
2. Formal verification of the specification after the translation of a CFSM into a finite-state-machine formalism;
3. Manual hardware/software partitioning. The granularity level is the CFSM;
4. Hardware/software co-simulation based on the Ptolemy simulation framework [4]. Related to the hardware/software partitioning, the microprocessor selection, and the scheduler selection, the architectural

tradeoffs are explored and evaluated, relying on code size and performance estimates of the processor;

5. Hardware synthesis of the CFSM sub-network by mapping into the BLIF (Berkeley Logic Intermediate Format) format. Each transition function is a combinational circuit, optimized by logic synthesis techniques, where states variables are implemented by registers. An XNF netlist can be generated to get a FPGA Xilinx prototype;

6. Software synthesis of the CFSMs sub-network into a C code structure which includes one procedure for each CFSM and a real-time operating system;

7. Synthesis of the interfaces between the different implementation domains: hardware, software, and the environment.

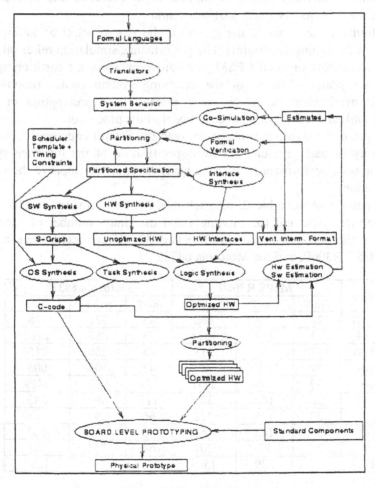

Figure 3. The POLIS system

5. ARCHITECTURE EXPLORATION OF THE HARDWARE/SOFTWARE DESIGN SPACE

We modified the Esterel specification of the ARINC interface to use it as a front-end language in the POLIS system, as regards to its asynchronous communication model. We altered the code of some CFSMs that react to events coming from at least two distinct modules in order to get a correct scheduling. The Ptolemy graphical interface enables to interconnect CFSMs before functional simulation.

POLIS/Ptolemy provides to the user a rich library of components for the test bench. Many simulation scenari were used in the Ptolemy debug mode. The ARINC interface system with four input channels and two storage indexes was verified by functional simulation.

Performance analysis of the system is the key to select an architecture that meets the timing constraints. The performance simulation relies on the C generated models of each CFSM, the hardware/ software partitioning, the scheduling policy chosen for the operating system (static round-robin, dynamic mechanism based on static priority with preemption or non-preemption), and the timing and cost model of the processor.

We noted the significant impact of both the Esterel code quality and the granularity of each module. An over-specification of the interface system results in poor performance. We used many iterations to improve the system specification.

We present in the table 2, for each module, the estimated results of the code size in bytes, and the minimum and maximum number of execution cycles of the selected processor. The two 32-bit microcontrollers are the MIPS RISC R3000, and the Motorola 68332.

Modules	MIPS R3000			Motorola 68332		
	minT	maxT	code	minT	maxT	code
Gap (x4)	27	154	400	43	371	328
Msg (x4)	33	424	1314	73	867	1374
Req (x4)	27	78	283	43	286	215
Scanner	27	95	732	43	303	603
Arbiter	39	495	2237	124	953	2419
Acq	38	162	326	112	496	243
Date	28	125	222	66	290	163
Raf (x2)	27	148	452	43	379	367
Write	27	83	207	64	277	161
Read	27	229	1508	43	745	1298

Table 2: Estimated results for a 4-input channel interface: code size (in bytes) and execution times (in clock cycles)

The system must both perform the data acquisition of each input channel at a rate of 10 µs, and respond to the asynchronous read request at a minimum interval of 6 µs. The hard timing limit for the return of a data is 3 µs. We performed worst case simulations, i.e. with a maximum rate of true ARINC packets received concurrently over all the channels.

An architecture that consists of the modules controlling the reception of ARINC packets over each channel (called GAP, MSG, and REQ) mapped to hardware, and the others mapped to software can meet the timing constraints. A 4-channel interface requires a R3000 frequency of 10 MHz. An 8-channel interface needs 25 MHz, and a 16-channel needs 50 MHz. The final 56-channel ARINC interface could be implemented with the same partitioning and scheduling with a R3000 at 100 MHz.

6.   CONCLUSION AND FUTURE WORK

This paper presented the hardware/software codesign of an industrial example: the avionics ARINC interface system. The first step of our work was to get a system-level executable specification, abstracting any architecture-dependent detail. The synchronous Esterel modules, slightly modified, were used in the globally asynchronous model CFSM of POLIS for distributed systems.

The POLIS system is convenient for critical control-dominated systems. The future work will consist of extending the POLIS library with another microcontroller model with better performance like the Power PC. Finally, co-synthesis and prototyping of the full ARINC interface system with the POLIS codesign flow represents another work axis.

7.   ACKNOLEDGMENTS

We would like to thank all the members of the Esterel team and especially the POLIS team for their precious help.

8.   REFERENCES

[1] F. Balarin, M. Chiodo, P. Giusto, H. Hsieh, A. Jurecska, L. Lavagno, C. Passerone, A. Sangiovanni-Vincentelli, E. Sentovich, K. Suzuki, et B. Tabbara. *Hardware-Software*

Co-Design of Embedded Systems, The POLIS approach, Kluwer Academic Publishers, 1997.

[2] F. Balarin, L. Lavagno, P. Murthy, A. Sangiovanni-Vincentelli. "Scheduling for Embedded Real-Time Systems" *IEEE Design &Test of Computers*, pp.71-82, Jan. 1998.

[3] G. Berry, G. Gonthier. "The Esterel Synchronous Programming Language: Design, Semantics, Implementation", *Science of Computer Programming* Vol. 19, N°2, pp. 87-152, 1992.

[4] J. Buck, S. Ha, E.A. Lee, and D.G. Masserschmitt. "Ptolemy: a framework for simulating and prototyping heterogeneous systems". *Int. Journal of Computer Simulation*, special issue on Simulation Software Development, Jan. 1990.

[5] R. Ernst. "Codesign of Embedded Systems: Status and Trends", *IEEE Design & Test of Computers*, pp.45-54, April 1998

[6] R. Ernst. "Target Architectures". in W. Wolf and J. Staunstrup *Hardware/Software Co-Design: Principles and Practice*. Kluwer Academic Publishers, 1997.

[7] D. Gajski, F. Vahid, S. Narayan, et J. Gong. *Specification and Design of Embedded Systems*, Prentice Hall, 1994.

[8] N. Halbwachs. *Synchronous Programming of Reactive Systems*, Kluwer Academic Publishers, 1993.

[9] N. Halbwachs, P. Caspi, and D. Pilaud. The Synchronous Dataflow Programming Language Lustre. *Another look at Real Time Programming, Proceedings of the IEEE*, Special Issue, Sept. 1991.

[10] P. Le Guernic, M. Le Borgne, T. Gauthier, and C. Le Maire. Programming Real-Time Applications with Signal. *Another look at Real Time Programming, Proceedings of the IEEE*, Sept. 1991.

[11] C.L. Liu and J.W. Layland. "Scheduling algorithms for multiprogramming in a hard-real-time environment" *Journal of the ACM*, Vol.20, N°1, pp. 46-61, Jan. 1973.

Using SDL to Model Reactive Embedded System in a Co-Design Environment

Rajesh Kumar

Cadence Design Systems (I) Pvt. Ltd. 57A & B, NEPZ, Noida-201305, UP, India
Ph: +91-11-91-562842, Fax: +91-11-91-562231, email: rajeshk@cadence.com

Key words: SDL, CFSM, HW/SW Co-Design

Abstract: Specification of the system behavior in a high level language is an important task in a hw/sw co-design environment. To simplify the design process one should be able to use the same high level language for specification, simulation as well as synthesis of the system. In this paper we present an approach to model the system behavior using a subset of Specification and Description Language (SDL)[1, 2, 3, 4], and describe the scheme used to migrate the design specification in SDL to a hw/sw independent form of model of computation, called Co-Design Finite State Machine (CFSM)[5, 6]. CFSM based model of computation has been used in an upcoming co-design solution from Cadence for embedded system design, simulation and synthesis. Our SDL subset can also be simulated using existing SDL simulators in the market, thus achieving a seamless integration between new codesign solution and the existing SDL simulators.

1. Introduction

Embedded systems for reactive applications are implemented as mixed hw/sw systems. Methods for designing embedded systems often require to specify and design hardware and software separately. A specification, often incomplete and written in non-formal languages, is developed and sent to the hardware and software engineers. Hw/sw partition is decided a priori and is adhered to as much as possible, because any changes in this partition may necessitate extensive redesign. Designers often strive to make everything fit in software, and off-load only some parts of the design to hardware to meet timing constraints. The problems with such design methods are:

- Lack of a unified hw/sw representation, which leads to difficulties in verifying the entire system, and hence to incompatibilities across the HW/SW boundary.
- A priori definition of partitions, which leads to sub-optimal designs.
- Lack of a well-defined design flow, which makes specification revision difficult, and directly impactstime-to-market.

The proposed solution is centered around CFSM. A CFSM model of a system's behavior is a Network of Extended FSMs (EFSM). This representation allows us to preserve semantic correctness throughout the design process, because CFSMs assume unbounded, non-zero reaction delays, that correspond both to hardware and software behavior. The CFSM model is not meant to be used directly by designers, due to its relatively low level view of the world. Designers will conceivably write the system specifications in a higher level language that will be directly translated into

121

J. Mermet (ed.), Electronic Chips & Systems Design Languages, 121–130.

CFSMs

SDL is a higher level language specified in ITU-T Recommendation Z.100. It is well suited for describing stimuli-response behavior since it is based on the experience of describing telecommunications systems functions as communicating finite state machines. The core of an SDL system is an SDL process which is an EFSM.

Both CFSM and SDL process being based on EFSM, there are a number of behavioral similarities between them. Barring some differences in their communication and scheduling semantics, SDL is a good language for authoring and simulating CFSM network. This paper presents a scheme to overcome these differences and identifies the synthesizable subset of SDL for specification and simulation of a network of CFSMs in an SDL simulator. The paper also suggests ways to simulate effects of various scheduling/implementation schemes (e.g. Round Robin RTOS scheduling or implementation in H/W using a clock tick etc.) on the behavior of a CFSM.

SDL supports a number of constructs which can be used to describe an EFSM in an SDL process diagram. With these constructs it is fairly easy to describe the behavior of a CFSM. But when it comes to communication semantics there are some differences between a CFSM and an SDL process. To overcome these differences we need a layer around the CFSM behavior which can exhibit SDL communication semantics on the outside and at the same time provide a CFSM like environment to the enclosing CFSM behavior. In the proposed solution, we have achieved this by modeling a CFSM as a dual service SDL process.

The dual service SDL process which has been used to model the CFSM, can co-exist with any other SDL process in a design. It can be interconnected with other CFSM/non-CFSM SDL processes and the complete system can be simulated using any SDL simulator.

This paper is organized as follows. In section 2, we give an introduction to CFSM and its behavioral semantics. Section 3 gives an overview of the SDL language and presents some of its important constructs, syntax and semantics, which are relevant for specifying the behavior of a CFSM. Section 4 presents the differences in the communication semantics of CFSM and SDL, and it introduces our proposed solution. With the help of an example, section 5 gives details of the solution, how to model the CFSM using our proposed scheme, and how does this solution preserve CFSM semantics in SDL. Section 6 presents some ideas of future work in this direction. In the end, section 7 concludes the paper and outlines various applications of the proposed solution in HW/SW co-design.

2. CFSM

A Co-design Finite State Machine (CFSM), like a classical Finite State Machine, transforms a set of inputs into a set of outputs with only a finite amount of internal state. The difference between the two models is that the synchronous communication model of classical concurrent FSMs is replaced in the CFSM model by a finite, non-zero, unbounded reaction time. This model of computation can also be described as Globally Asynchronous, Locally Synchronous. Each element of a network of CFSMs describes a component of the system to be modeled. The CFSM specification is apriori unbiased towards a hardware or software implementation. While both perform the same computation for each CFSM transition, hardware and software exhibit different delay characteristics. A synchronous hardware implementation of CFSM can execute a transition in 1 clock cycle, while a software implementation will require more than 1 clock cycle.

The execution of a CFSM is invoked by an external agent (e.g. the scheduler of a RTOS if the CFSM is implemented in SW). When a CFSM is invoked for execution, if the input stimulus matches one of the specified stimuli, the CFSM makes a transition and consumes all the events that are present at the time of invocation. It may possibly make the *trivial transition* (i.e. the input events are consumed but no output events are emitted and no output values or states are set). If the input stimulus does not intersect any specified transition, the input events will NOT be consumed; this is called an *empty execution*.

A CFSM sends events to other CFSM which may detect them when they are executed. Pending events are stored in buffers of size 1.

The sequence of output events produced by a CFSM in response to a sequence of input events is determined by the CFSM behavior and by the scheduling i.e. the sequence of executions of this CFSM and other components. No scheduling constraint is imposed apriori. This and the fact that the communication occurs over queues of fixed size, makes it is possible that some occurrence of an event is missed by a receiver CFSM.

3. SDL

SDL is a formal, graphical, object oriented language. It is suitable for describing reactive, discrete systems. Where a *reactive system* is a system whose behavior is dominated by interactions between actions input to the system, and the reactions output by the system. And a *discrete system* is a system whose interaction appears at discrete points and by means of discrete events.

An SDL system has a set of *blocks*. Blocks are connected to each other and to the environment by *channels*. Within each block there are one or more *processes*. These processes communicate with one another by asynchronous *signals* and are assumed to execute concurrently.

The behavior of an SDL process is described as an EFSM. When started, a process executes its *start transition* and enters the first *state*. The reception of a signal triggers a *transition* from one state to a next state. In transitions, a process may execute *actions*. Action can assign values to variable attributes of the process, branch on values of expressions, call procedures, create new process instances and send signals to other processes.

A process can also be specified in terms of a composition of *services* each with its own EFSM. Services are useful when the behavior of a process can be described as a number of independent activities (only sharing data). When services can be successfully applied, they can reduce the number of states in a behavior description considerably.

Services in one process execute one at a time, that is not concurrently. When the executing service reaches a state, the service capable of consuming the next signal in the input port of the process instance takes over execution.

The input port of a process instance is an unbounded FIFO-queue. SDL semantics assume that all the input signals to a process are stored in a single FIFO queue and the process takes a signal from the head of the FIFO to make a state transition in the process.

Parameterized types is a generalization mechanism in SDL that makes type definition partly independent of where they are localized. Parameterized type definition can be made independent of enclosing scope by means of *context parameters*. For example by putting appropriate context parameters with desired constraints (properties) a service type can be written which can be analyzed independently without instantiating it in a process.

In some cases it is convenient to express that reception of a signal takes priority over reception of other signals. This can be expressed by means of *priority input*.

4. CFSM specified as an SDL Process

With constructs like states, signal input/output, data variables, conditions, assignments etc., behavior of a CFSM can be easily specified in SDL. But there are some differences between communication semantics of a CFSM and an SDL process.

- **Event Loss in CFSM vs. Unbounded input Queue in SDL**: In CFSM, if more than one event occurs on some input port before the CFSM is scheduled for execution, then only the last event is preserved. This is because a 'one-place-buffer' is used for storing the input events. In SDL, however, an unbounded queue is used for storing input events. So all event occurrences are preserved till the SDL process consumes those events.
- **Partial ordering of input events in CFSM vs. Single FIFO ordering for all input events in SDL**: If two events (on two different input ports) are 'present' when a CFSM is scheduled for execution then the CFSM doesn't have any information regarding their relative order of arrival. On the other hand, In an SDL process all the events (on the same input or on different input ports) are put into a SINGLE FIFO queue. So even if two events have arrived simultaneously they will be placed in the queue one after the other and hence will be processed sequentially, one at a time.

In the proposed solution, we have overcome these differences by modeling a CFSM as a dual service SDL process. The SDL process is divided into two SDL services, the *control service* which models the inputs and the scheduling part of the CFSM, and the *behavior service* which models the transition relation (the behavior) of the CFSM.

The CFSM input ports are mapped to SDL signals which are connected to the *control service*. One-place-buffer for each input port is modeled as a local variable in our SDL process. When an event occurs on any of the input signals, the *control service* updates the value of the corresponding local variable. Scheduling of the CFSM is modeled using a special input signal, Tick, which is used to direct the *control service* to pass control to the *behavior service*. *Behavior service* uses the current value in the one-place-buffers (local variables in our SDL process) to do the appropriate state transition.

Following our scheme, the user will be required to create the *behavior service* part of the CFSM only, which will be in the form of a state transition diagram. The *control service* and rest of the SDL process can be generated automatically. For entering the *behavior service* we have identified a subset of SDL which can be used to specify both the data as well as the control part of the CFSM. Interface of the CFSM is specified as *context parameters* (An SDL feature which is used in writing parameterized service types etc.) of the *behavior service*. This helped us to make the *behavior service* a self-contained entity which can be compiled independently.

Also note that our objective here is to provide a convenient way to model a CFSM in SDL. So we shall select only a subset of SDL which is useful in describing the behavior of a CFSM. As a result some features of SDL like dynamic process creation etc. will not be supported.

A CFSM, as opposed to a pure explicit FSM, is an efficient way to represent and implement sequential behavior because the behavior specification is extended with the use of *side functions* such as arithmetic operations that absorb some of the complexity of an FSMs transition relation. So we shall support SDL constructs like *variables, expressions, data types, literals, operators* etc. also, which can be used to specifying complex expression in a CFSM.

5. Dual service SDL Process

To explain how to model a CFSM in our scheme we shall make use of an *ExampleCFSM*. It is a fully contrived example in order to capture various aspects of CFSM specification in as simple form as possible.

Input Ports
A: Pure event
B: Pure value of type integer
C: Event with value of type integer
Output Ports
O1: Pure event
O2: Event with value of type integer
Behavior
There are three cases to be handled:
1. If both A and C are ON: Look at the value of "B+C", if it is less than 5 then output the signal O1, if it is between 5 and 8 then output the signal O2 (with value 10), otherwise output both O1 as well as O2 (with value 20).
2. If A is ON but C is OFF: Wait till C is also ON and output O1.
3. If A is OFF but C is ON: Wait till both A and C are ON and output O2 (with value "B+C").

The CFSM maps to a process in SDL. This process has two important parts the *Behavior service* and the *Control service*. The *Control service* handles all the inputs to this process and gives control to the *Behavior service* as and when the inputs need to be processed to produce the outputs.

Each of the CFSM input port maps to a separate SDL input signal [signal A, B(integer), C(integer)]. These SDL signals will be referred to as *CFSM_IN signals* from now on. Corresponding to each one of the valued CFSM_IN signal, there is a local variable for storing the value of the signal [dcl B.v, C.v integer]. These variables will be referred to as *CFSM_IN value variables* in rest of this document. Similarly for every CFSM_IN signal except PureValues, there is a local variable for storing the presence/absence status of the signal [dcl A.p, C.p Boolean]. These variables will be referred to as *CFSM_IN presence variables* in rest of this document.

The CFSM output ports map to SDL output signals [signal O1, O2(integer)]. These will be referred to as *CFSM_OUT signals* in rest of this document.

There is an input signal *Tick* which can be used by an external scheduler to schedule the execution of this CFSM.

In addition to these there are two internal signals, *Execute* and *Done*. These are used between *Control service* and the *Behavior service* to exchange control of execution. There is an internal variable, *dcl EmptyExecution Boolean* which is used to distinguish an empty execution from a normal execution.

Its SDL/GR looks like the following:

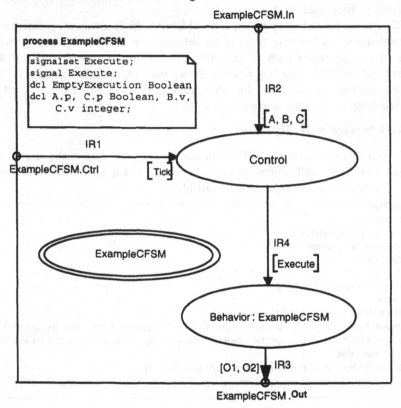

Let us now see how does this work. Whenever an event arrives at the CFSM_IN signals the control service updates its corresponding CFSM_IN variables (visible to the behavior service also). When it decides to execute the CFSM (more on this later) it sends the *Execute* signal to the behavior service and gives the control to the behavior service. When the control comes back from the behavior service then the control service resets the corresponding CFSM_IN presence variables. In case of empty execution the behavior service is expected to set a variable *EmptyExecution* to value *TRUE* which directs the control service not to reset the CFSM_IN presence variables.

5.1 Behavior Service

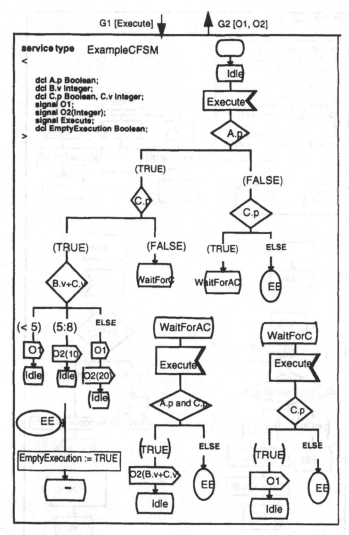

This is the part of the process which the user inputs. The behavior service gets the control through signal *Execute*. It then executes certain decision statements and outputs zero or more CFSM_OUT signals before changing its state. In case of empty execution it doesn't output any signal and doesn't change state, it just assign *TRUE* to the variable *EmptyExecution* and remains in the same state.

5.2 Control Service

This part of the process can be generated automatically by looking at the behavior service. The control service has two important responsibilities, first it handles the CFSM_IN inputs coming to this process in the form of SDL signals, and second it handles scheduling/execution of the behavior service.

128

Its SDL/GR looks like the following:

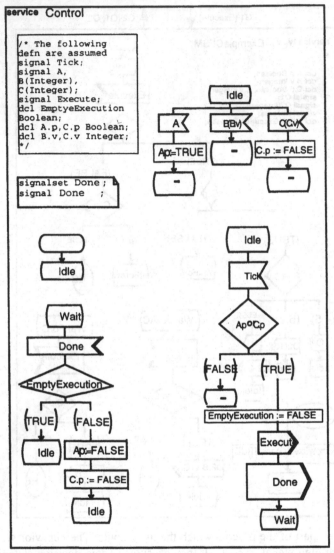

Whenever a CFSM input arrives the control service updates the corresponding local variables. Directive to execute the behavior comes from outside in the form of the input signal *Tick*. On receiving *Tick* the control service checks whether any of the CFSM inputs have an event present or not (only PureEvent and ValuedEvent variables are checked). If no event is present then the control is returned back without executing the behavior. If one or more events are present then the behavior is executed by sending the signal *Execute*. At the same time the control service sends a signal *Done* to itself so as to resume control after behavior has finished execution.

Note that the order in which *Execute* and *Done* are output, is important because *Control Service* and *Behavior Service* being part of the same SDL process share the signal FIFO Q. So first the signal *Execute* will trigger the *Behavior Service* and immediately after that *Done* will automatically trigger the *Control Service*. This scheme makes *Behavior Service* easier to write because the *Behavior Service* does not have to send an explicit *Done* to the *Control Service*. Also note that *Execute* and *Done* are *priority* inputs, as SDL semantics ensure that priority inputs will be processed before other inputs so it is guaranteed that the *Control Service* gets back control as soon as the *Behavior Service* is done with its execution.

On getting the control back the *Control Service* resets the presence status of the CFSM_IN presence variables depending upon the value of the variable *EmptyExecution* (initialized to *FALSE* before executing behavior, set to *TRUE* by the behavior service in case it wants to go through an empty execution).

5.3 Ensuring the CFSM Semantics

Let us now see, how does this SDL process satisfies the CFSM semantics:

- **Input Events:** Input events are stored in a buffer (local variable) of size 1 till the time they are consumed by executing the behavior service of the process.

  Event loss of CFSM is modeled by over-writing of the CFSM_IN variables if some CFSM input arrives more than once before the CFSM is actually scheduled for execution (i.e. the SDL process receives a *Tick*).

  Using this scheme of CFSM_IN variables the *Behavior Service* get to see (and consume) events on all its input ports simultaneously. This is in contrast to the normal one-at-a-time input handling of SDL (because of the single FIFO-Q for all input ports).

  Pure valued inputs are also maintained as a local variable which is nothing but a continuous up-to-date value made from the discrete events arriving at the corresponding CFSM_IN signal. This strategy of implementing continuous value exchange in SDL using discrete signal communication semantics is similar to the one documented in the Z.100 document - *SDL Methodology Guidelines*.

- **Empty Execution:** Empty execution is supported by using the variable EmptyExecution which is normally set to FALSE and leads to resetting of the presence status of various CFSM_IN presence variables. But in case of empty execution it is set to TRUE and ensures that the CFSM_IN presence variables are not reset.

- **Scheduling**: The input Tick provides external control to the scheduling of the CFSM process. The Control service executes the behavior only when it receives the Tick.

- **Signal Output:** Unlike signal input the signal output doesn't go through the Control service. It is plane and simple SDL signal output directly from the Behavior service. Reason being that the CFSM semantics are ensured by the receiver of these signals and the sender doesn't have to do anything special for that.

6. Future Work

Future direction for this approach is to extend the SDL subset presented in this paper. This can be achieved by identifying the SDL constructs which cannot be done in hardware or are not opimized if done in hardware (e.g. dynamic process creation etc.) and separate those portions of design as software only portions. Rest of the design can follow the normal hw/sw co-design approach. Another extension possible is to support the input signal queue of length more than one. This can be done by placing pre-synthesized implementations (both software as well as hardware versions) of the queue as library elements and to choose the appropriate implementation while synthesizing the system.

7. Conclusions

SDL is a good choice for modeling a CFSM because it is based on EFSM and provides a wide variety of language features to specify an extended finite state machine. It is a formal specification language, at the same time it is very user friendly and supports a graphical flow chart like notation which simplifies the task of specifying a CFSM behavior. SDL has been used extensively in the telecommunication application design. As a result a number of tools exist today which support various types of analysis e.g. simulation, formal verification etc. for an SDL description.

Capability to model a CFSM in SDL gives us the benefit of high level specification and verification of our design. And at the same time it provides us an efficient way of doing estimation, simulation, synthesis and hw/sw codesign of our system using CFSM model.

This work is part of a complete HW/SW co-design solution being developed in Cadence.

References

1. A. Olsen, O. Faergemand, B.Moller-Pedersen, R.Reed, J.R.W. Smit. Systems Engineering Using SDL-92. North-Holland, 1994.
2. [ITU Z.100 SDL-92] ITU, Geneva. Specification and Description Language (SDL), 1994
3. [ITU Z.100 app. I] ITU, Geneva. SDL Methodology Guidelines, 1994.
4. [ITU Z.100 annex F] ITU, Geneva. SDL Formal Definition, 1994.
5. M. Chiodo, P. Giusto, H. Hsieh, A. Jurecska, L. Lavagno, A. Sangiovanni-Vincentelli. A Formal Specification Model for Hardware/Software Codesign. In Proceeding of International Workshop on Hardware-Software Codesign, October 1993.
6. F. Balarin, M. Chiodo, P. Giusto, H. Hsieh, A, Jurecska, L. Lavagno, C. Passerone, A. Sangiovanni-Vincentelli, E. Sentovich, K. Suzuki, B. Tabbara. Hardware-Software Co-Design of Embedded Systems: The Polis Approach. Kluwer Academic Press , June 1997.

A Synchronous Object-Oriented Design Flow for Embedded Applications

P.G. Plöger, Reinhard Budde, Karl H. Sylla[1]
GMD-AiS, Schloß Birlinghoven, D-53754 Sankt Augustin, Germany
{ploeger,reinhard.budde,sylla}@gmd.de

Abstract The selection of hardware components for embedded controllers is strongly influenced by timing constraints to be met. We present a design flow for synchronous object-oriented for embedded applications, which integrates timing estimations into the design process. It is based on the language synchronousEifel *sE* which unifies the synchrony hypothesis with OO design principles. *sE* uses Synchronous Automatons, a compact and optimized intermediate language. This representation allows to apply both hardware evaluation and software optimization techniques. Code is produced in a retargetable way which enables execution time evaluation on a high level. Thus the synchrony hypothesis may get validated, target processor selection can be postponed to a late design stage, and true hardware-software co-design becomes possible. An example for a successful design is given.

1 INTRODUCTION

From our experience the following questions need to be addressed in an embedded system design flow:

- how to get real-time aware, verifiable and reusable specifications?
- how to generate the complete system implementation obeying timing constraints from this specification?
- how to enable late design decisions (e.g. selection of a μ-Controller as target machine) without sacrificing too much of the early design efforts?

To all this points there are partial answers. Many techniques are available to specify and to visualize the *dynamics* of a system. Examples are Petri nets [17], message sequence diagrams [11], or annotated FSM based models like hierarchical FSMs [10], FSMs with data path [7] and co-design FSMs [15]. Such models are a good basis for simulation and experimentation with the system, but they fall short in precise modeling the real-time execution behavior

---

[1]Supported by BMBF Contract 01M3035A (ABS) and Esprit LTR-Project 22703 (SYRF)

J. Mermet (ed.), Electronic Chips & Systems Design Languages, 131–142.
© 2001 *Kluwer Academic Publishers.*

of a system. Furthermore, if real-time properties are specified and used for validation and test, this can only achieved on a fairly low level of abstraction.

A viable abstraction of the concrete notion of time is achieved by the synchronous approach. It is based on the *perfect synchronization hypothesis* which assumes that the system reacts to external events and that any reaction to an event is finished before the next event may occur. The reaction itself is atomic and while processing it cannot be influenced by the environment. Advantages of this hypothesis are: It defines a simple and understandable execution model, which is familiar to many control engineers; it enforces a deterministic and reproducible behavior of the program itself. Languages based on this execution model are Esterel and Argos, well suited to specify control-dominated applications, or Lustre, which is able to capture data flow (see [2]). Programs written in the formalism of StateCharts can be understood in a similar way.[10]

But in the synchronous setting the complete design flow from high-level synchronous specification to real world implementation is not well supported. Our approach addresses this problem. This includes for hard real-time systems a reliable analysis, that the synchronous hypothesis is valid. This is done by computing worst case reaction time based on the generated code, parameterized by target microcontroller and C compiler. Thus hardware selection can be postponed after all complex design decisions are taken. The basis for this computation are Synchronous Automatons syA, which are used [18] as a common semantical model for synchronous languages like Esterel, Lustre, Argos[2], and sE[4].

Further support of an established methodology to allow *reuse* is missing. Object oriented software approaches [3] meet this challenge. Furthermore they can cope with the increasing complexity of embedded systems software. The implementation under real-time constraints is an aspect usually not directly supported in OO systems. This problem has been tackled by a number of researchers, though. Systems like Adept[13], Cathedral[21], Castle[19], Chinook[5], Polis[1], Cosmos[12], Cosyma[6], Lycos[16] or Vulcan[9] allow specification of timing bounds to generate complete hardware and software implementations from a specification. They focus on HW/SW co-design with its questions of partitioning, HW selection or production, SW and communication synthesis. The problem tackled in this paper is the integration of object-oriented design on the base of synchronous specification and automated performance estimation.

This paper is organized as follows: after stating the underlying principles of the design flow in section 2, we elaborate on code generation and timing estimation in section 3. Here we show how tools for real-time analysis and partitioning apply to the generated code. This enables the analysis of real-time behavior with respect to different target architectures and processors. The example given in section 4 is an excerpt of industrial embedded systems developments and summarizes the advantages of our approach.

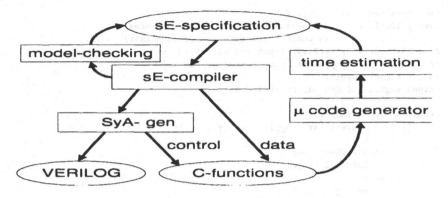

Figure 1: the overall design flow

2 FOUNDATION OF SE DESIGN

Our design flow is depicted in Figure 1 and consists of the following (possibly iterative) steps:

- system definition by sE classes, which encapsulate both real-time behavior and data. The application is a tree-like structure of objects of these classes, rooted in one instance of a configuration class, which is responsible for initialization.
- analysis and validation, including model-checking, by the compiler. This uses Synchronous Automatons syA as intermediate language.
- C code generation, based on syAs.
- target system code production, based on the C code.
- timing constraint check and target hardware selection.

Note, that validation, including model-checking, is done on the same intermediate language as code generation, i.e. no separate model has to be constructed for validation. This increases substantially the designer's belief, that the application fulfills proven assertions.

2.1 A Simple Example

A simple application is given in Figure 2. It shows a class definition, that switches LED's on and off. The first part defines timing requirement and input and output signals, the reactive interface. The implemetation of the reactive ("real-time sensitive") interface is done in the reactive method prefixed by the keyword **reactive**. Methods can be called and fields accessed.

Finally a class may contain a data part, e.g. the field (attribute, data member) **flagRed** and the method (procedure, member function) **doToggle**. This part is built according to the object-oriented concepts and design styles.

In sE it is encouraged to specify permissible scheduling for data usage. This expresses invariants the designer has in his or her mind and avoids time races (non-determinism). In the example only schedules are legal, where **doToggle**

```
configuration class Toggle {
timing 10millissec; \\ mimimal temporal distance between
                    \\ consecutive activations (instants)
// control part: interface: input and output signals
//               implementation: a reactive method
input  signal toggle;
output signal red_led_on, green_led_on;

reactive toggle_leds {
    [[ loop { await next toggle; doToggle(); };
    || loop {
        if (flagRed) then
            emit red_led_on;
        else
            emit green_led_on;
    };
    ]];
}

// data part: private, only accessible for control part
boolean flagRed;
void doToggle() { flagRed = !flagRed; }

// constraints for scheduling: in all instants
// doToggle must be called before reading flagRed
sequence change_before_read: doToggle < flagRed;

} //end class Toggle
```

Figure 2: Simple example for a *sE* class with control and data part

is executed in any case before flagRed is read (< reads "always before"). If the reactive implementation and the sequence specification do not gurantee a deterministic schedule of data usage (a potential time race) this will be detected by the *sE* compiler based on data flow analysis and the program will not compile.

2.2   The Synchronous Paradigm

In our synchronous framework a reactive program has to respond to stimuli (signals) from the environment. It consists of objects. Some of the objects observe signals and emit signals; these are called reactive objects. The behavior of reactive objects is modeled using the synchronous approach.

A synchronous system does not react continuously to the environment. Two phases are distinguished: the system is reacting or it is idle. Signals on input lines are always collected, but never propagated immediately to the system. The change from the idle phase to the reacting phase is effected by an activation (hardware designer would call this a clock pulse). Whether this is done periodically, using a timer subsystem of a microcontroller, or depending on environment conditions is irrelevant for this discussion. The only requirement is, that between activations at least the time requested in the timing-clause passes. When the activation takes place, the gathered signals make up the input event and are provided to the system.

Now the program is reacting and computes the response. Signals are emitted or tested whether they are present. Subreactions are computed in parallel. The set of output signals emitted during the reaction is called the output event. The reaction is complete, if all components of the (non-sequential) system have committed to halt. Then all signals of the output event are made available to the environment and the program is idle again.[2]

Usually output signals are connected to actuators or to display units and effect changes in the environment. These changes may produce input signals, which are collected. Then, if the clock ticks, the next reaction step is initiated.

Each step of input→reaction→output is called an instant. While reacting the system is logically disconnected from the environment, i.e. it is impossible to add input signals to the input event during a reaction. This model may appear simple and natural to a hardware-designer. But it is unusual in software design. This idea of atomicity of reaction applies to data, too: also calls to methods are atomic and cannot be interrupted by incoming signals (they are stored for the next reaction). This is a powerful paradigm, as it allows to design nonsequential systems with deterministic, reproducable behavior (compare this with potential time races in threaded systems, when threads compete w.r.t. system resources. Thus this paradign is well-suited for dependable systems.

But to make use of the full power of the paradigm, on the other hand one has to prove, of course, that if a real-time requirement is given (e.g. the system has to sample data exactly at 9,600 Hz) the system will have finished a reaction to an input event in *any* case before the next input event arrives. This is proven (later) with respect to selected hardware and compilers (see section 3). The property that a system is always ready to react to an input event when the environment requests that is called *perfect synchrony*.

Further there is no rendezvous-like concept in the synchronous model. Emitting a signal will never be blocked. This would be the case in asynchronous languages, where emitters (senders) are blocked if no process (receiver) awaits the signal. Further, signals are not dedicated to a specific receiver, they are broadcasted to all components of the program, and they are not consumed by the await statements. This facilitates simultaneous reactions to the same signal.

The synchronous model enables us to apply hardware evaluation techniques to the real-time behavior of the whole application. From an operating system point of view the generated target-code can be considered as a generated static scheduler for calling methods and accessing fields of the data type, where the activation conditions are boolean formulas made from signal and state information.

---

[2]To use the processor during the time in which the systems gathers signals only, i.e. the reactive part is passive, time uncritical computations may be scheduled. No assumption on the execution time of such activities can be made. This is the reactive part for.

2.3   The Synchronous Automaton

We define synchronous behaviour in terms of a particular kind of state machine, we refer to as *Synchronous Automata (syA)*.

Definition 1 *A Synchronous Automaton $\mathcal{P}$ is represented by a tuple*

$$(\mathcal{S}, \mathcal{I}, \mathcal{O}, \mathcal{R}, \alpha, \mathcal{P}^{\rightharpoondown}, \mathcal{P}^!)$$

where

- $\mathcal{S}$ *is a set of* signals *with* $\mathcal{I}, \mathcal{O} \subseteq \mathcal{S}$ *being sets of* input *and* output *signals,* $\mathcal{R}$ *is a set of* registers, *$\alpha \notin \mathcal{R}$ is a special register representing the initial state when true, all other states when false; and where*
- $\mathcal{P}^{\rightharpoondown} : \alpha \times St \times 2^{\mathcal{S}} \to St$ *is a transition function, and* $\mathcal{P}^! : \alpha \times St \times 2^{\mathcal{S}} \to 2^{\mathcal{S}}$ *is an output function.* $St = 2^{\mathcal{R}}$ *is the set of states.*

Here $2^{\mathcal{X}}$ is the power set of $\mathcal{X}$. The synchronous automaton is specified in terms of several kinds of functions:

- for each signal $s \in \mathcal{S}$, a *presence function* $\delta(s) : \alpha \times 2^{\mathcal{S}} \times 2^{\mathcal{R}} \to 2^{\mathcal{S}}$ specified by a boolean equation of the form $s = \phi$,
- for each register $r \in \mathcal{R}$, an *activation function* $\delta(r) : \alpha \times 2^{\mathcal{S}} \times 2^{\mathcal{R}} \to 2^{\mathcal{R}}$ specified by a boolean equation of the form $r = \phi$,

$\phi \in B(\mathcal{S} + \mathcal{R} + \alpha)$ where $B(X)$ is the set of boolean formulas over X, i.e., formulas built from elements in X using $\wedge, \vee, \neg$ as well as *true* and *false*. As an example consider the reactive definition:

```
loop
    await next i; emit s;
end;
```

i is input signal, s is output signal. The **await**-statement introduces a register r, because it has to be remembered, whether the program waits for i to be present in an input event. Thus one equation for the register r defines the next state function, and one equations for the output signal s defines the reaction function.

The program is compiled into a *syA* $\mathcal{P}$ with $\mathcal{R} = \{r\}$, $\mathcal{I} = \{i\}$, $\mathcal{O} = \{s\}$:

$$\mathcal{P}^{\rightharpoondown} : r = \alpha \vee (r \wedge \neg i) \vee (r \wedge i)$$
$$\mathcal{P}^! : s = i \wedge r$$

The next state function $\mathcal{P}^{\rightharpoondown}$ consists out of three parts: (1) α when the loop is entered, (2) $r \wedge \neg i$ if the program stays waiting when i is not present, (3) $r \wedge i$ if the program got i in an instant, the wait terminated and the loop is re-executed. The next state function is later optimized to $r \leftarrow \alpha \vee r$.

$\mathcal{P}$ works as follows. Initially all the registers are set to false, α is set to true since $\{\alpha\}$ is the initial state. At the initial reaction no signal will be emitted. The next state is $\{r\}$ (α does not belong to any follower state by definition of $\mathcal{P}^{\rightharpoondown}$ since $\alpha \notin \mathcal{R}$). Then signal s will be emitted in any instant the input signal i is present. $\{r\}$ is its own successor state.

The program computes a unique reaction in a finite amount of time. Because of synchronous communication, outputs are immediately available to calculate more outputs and this process has to terminate for the automaton to be reactive. This is captured by the existence of a fixed-point for the function $\mathcal{P}^!$. Programs with a unique fixed-point are called causally correct. This is checked by sE and excludes "anomalies" like

- programs, that have no fix-point at all. This is the case for example, when the emission of a signal triggers a reaction, that cancels the component, that made the emission.
- programs, that have more than one fix-point. This is the case in the following example, in which the reaction could be {} (the empty set) and {$s1,s2$}. Consider, that if s1 is present, also s2 must be present, but if s1 is absent, s2 must be absent too:
  [[if s1 then emit s2; || if s2; then emit s1;]]

A syA only contains boolean expressions. Advantages are, that equations (1) can be processed by BDD systems (e.g. model checking), and (2) they can be linked, comparable to the linking of modules based on external names. Equations of the reaction function with the same left-hand-side are or-ed, all others are merged. The causality check has to be redone. Thus we can successfully circumvent the state explosion problem which burdens the building of cross products in other FSM formalisms.

But the distinctive advantage is that a syA can be expressed in virtually any hardware or software description language. Back-end drivers are available for Verilog, BLIF, DC and C. Using the first alternative the scheduler can be burned into an FPGA thus ensuring very fast response times. The last alternative maps to C code which in turn can be compiled for the microcontroller present in almost all embedded systems anyhow.

3   CODE GENERATION AND TIMING

Following the design flow so far, we modeled the system using an sE specification. Thus we were able to validate non functional aspects like causality of the system specification, and can give feedback on the feasibility of a design with respect to the reactive part. We now need a procedure to verify the synchronous hypothesis.

According to the design flow in Figure 1 the generated C program falls into two parts: (1) a function for the syA, which encapsulates the control flow similar to a static scheduler. (2) all methods originating from the data part in the sE classes, e.g. method $doToggle$ in Figure 2. Both parts are validated against the original assumption that no function evaluation takes longer then the minimal observable time span between to clock ticks. While it is difficult to give hard real-time bounds for arbitrary programs [14] this task here is facilitated by the overall structure of the sE program. This has also been observed by other researcher like Suzuki [20], who reports promising runtime prediction results for the implementation of a co-design finite state

pseudo function	C equivalent	assembly code	comment
SetFalse(s3);	s3 = 0;	bclr reg,bitoffset	clear bit
SetTrue(s3);	s3 = 1;	bset reg,bitoffset	set bit
IfTrue(s0,l7);	if(s0) goto l7;	brset reg,bitoffset,label	branch if set
IfFalse(s0,l7);	if(!s0) goto l7;	brclr reg,bitoffset,label	branch if clear
Test(s4);	if(s4){...}	brclr reg,bitoffset,lelse	guarded block

Table 1: Pseudo functions and their assembly code for HC05

machine on a specific controller. The syA code is loop free, contains only single assignments, simple branches and guarded function evaluations. The evaluation of the data part functions are not restricted though and can have e.g. branching control flow, data dependant loops and/or data dependant operation timings. For these unrestricted functions the timing analysis is only simplified by the fact that the current target microcontrollers (e.g. M68HC05/11/12, 80C51) all show deterministic behavior. Furthermore interrupt routines are only allowed to register a new event but not to treat it since this would contradict the verified reactive semantic of the specification.

Code generation for the syA uses about 10 pseudo functions. They are mapped via preprocessor macros to optimized assembly code. This minimizes the machine dependencies for the code generation.

For instance, for a target controller in the Motorola family, (see Table 1) signals and registers are mapped to boolean one-bit variables, which are mapped in turn to bit offsets in memory bytes. Special branching assembly operations work on them. This results in a very small overhead. This dedicated code generation applies just to the syA the data part is mapped to plain C by the compiler.

The timing tool evaluates the runtime costs for the C functions for different machines $m \in Mach$ and runtime libraries. It is driven by three tables and a control/data flow graph (CDFG) containing all functions f and all their contained basic blocks $b_f \in BB$. The first table is a machine library and accounts for the cylce count of each C operation $op_j \in OP("C")$ on the different candidate controllers. The second is an execution profile which captures the execution frequency of each branch in the control flow of the function. The last one accounts for external costs since it contains cycle time bounds for functions linked in from external libraries.

The idea is to use C as a machine independent generic assembly code. The timing is stored as an interval $t_{op_j,m} = [op_{j,min}, op_{j,typ}, op_{j,max}]$ of minimal, typical or maximal possible cycle counts for each operation op_j. This takes care of data dependent operation timings. All arithmetic operators are distinguished via bit-sizes of the participating operands, while the rest are control flow operations with no operand at all, e.g. "goto", "call", "return". The timing intervals $t_{op_j,m}$ are either measured with special stop watch programs or this information comes from the respective data book. A dynamic profile is produced during execution of the system. It contains the calling count for each function and the execution counts x_{b_f} for each basic block. A static profile of the CDFG reveals the distribution of instruction in a given basic block, namely the counts n_{op_j,b_f}, which denote the occurrence number of operation op_j in basic block b_f. Now a runtime bound $T_{f,m}$ for each function f on controller m is calculated by:

$$T_{f,m} = \sum_{b_f \in BB} x_{b_f} \cdot c_{b_f,m} \tag{1}$$

$$c_{b_f,m} = \sum_{op_j \in OP} n_{op_j,b_f} \cdot t_{op_j,m} \tag{2}$$

Since the $t_{op_j,m}$ are actually intervals, so are $c_{b_f,m}$ and $T_{f,m}$. So far we just accounted for the worst case data dependent operation runtimes, but not for the worst case control flow. Control flow may split at loops and/or conditions. Data dependent loop bounds are considered by storing a maximum possible x_{b_f} for each basic block. The case of branch constructs like "if-then-else" is handled by taking the larger $c_{b_f,m}$ of the two basic blocks for the branches.

The external cost table is used for external runtime libraries, e.g. the float data type is unsupported for small microcontrollers but still may be needed. It allows to specify runtime intervals for whole function calls.

The calculation of an upper timing bound reads in the machine library and the CDFG. It then iterates over all functions, looking up their respective call frequencies in the execution profile. Then it follows the control flow graph of the function body weighting each edge with the observed execution count. The nodes of the control flow graph are basic blocks which are traversed in a second independent sweep. In a basic block each operation node is weighted according to the cycle count found in the machine library. If an external function was called from the data flow in the basic block the external cost table is used. If the basic block contains an expression tree –like in the cases when evaluating the syA boolean expressions– the whole tree is traversed. No shortcut evaluation takes place. This ensures that a hard upper bound for a logical expression is generated. We can thus guarantee an upper bound runtime for the syA. This cannot be achieved for the data functions when data dependent branches of the control flow exist. Here the best we can do is to use the most costly successor in the control flow always.

The timing analysis closes the second loop in the design flow. Now we know, whether the synchronous hypothesis holds true for the generated implementation. In cases when it fails we can either try to modify the specification, accelerate the code or choose faster hardware where we have again two choices; either put the scheduler functionality into a dedicated device or choose a faster microcontroller. Since up to now the design efforts are machine independent and can therefore be mapped easily to different hardware.

Acceleration in the syA code is possible via all techniques from hardware design for the optimization of sequential circuits. A promising method is the improved sequential redundancy addition and removal technique [8] which showed

Machine	SUN IPC	SUN Sparc10	186	386sx	386dx	486dx
Processor	Weitek	TI	Intel	Intel	Intel	Intel
Mhz	25	33	5.8	25	20	33
OS	SUN OS	SUN OS	DOS	DOS	DOS	DOS
Compiler	gcc -O2	gcc -O2	bcc -1	bcc -3	bcc -3	bcc -3
Timer method	getrusage	getrusage	8253 chip	8253 ...	8253 ...	8253 ...
Timer resolution	10^{-3}	10^{-3}	10^{-6}	10^{-6}	10^{-6}	10^{-6}
Time prediction	2.26	1.26	46.88	15.36	18.49	2.39
Time measured	2.28	1.01	38.04	5.65	8.4	1.33
Loop overhead	0.02	0.04	0.27	0.03	0.05	0.01

Table 2: Technical data of target processors, timing analysis results, measurements and overhead costs. Times given in seconds for 10000 iterations.

faster evaluation of up to 20%. A pure SW technique either exploits the single static assignment structure of the syA code and uses compiler optimization techniques, or hooks in at the code generation during the binding of bit variables to memory bytes. The set of boolean equations shows instruction level parallelism. For instance, one may search equal expression trees and try to bundle bit operands in bytes in such a way that the original expression is performed on the byte instead of a repeated expression evaluation working in bits. Because of the deterministic way how sE-programs are mapped to FSMs and because of common idioms there is strong evidence that this occurs frequently. The problem can be mapped to a question in paired register allocation. First trials show an acceleration potential in the same order of the redundancy removal technique.

4 RESULTS

Table 2 shows the results of the timing tool for a common example running on different architectures and a comparison of measured runtimes. The machines range from SPARC down to 16 Bit x86 based running under two different operating systems. This fact also determines the achievable precision of the timing tools. Since we are in a single user mode on a PC and have direct uninterrupted access to the timing HW we get a timer resolution which is finer by a factor of 1000. To address the coarse grained timers the evaluation of the synchronous automaton together with its data functions is repeated for 10000 times. The overhead for this repetition loop is indicated in the last row. In all cases it is marginal. Although it overestimated the runtimes for two architectures grossly, we observe that for all machines the timing tool predicted correctly the fulfillment of the synchronous hypothesis. Thus it was justified to base the design on it and we are still free to choose amongst some candidates for the final realization of the embedded system.

5 CONCLUSION

The design flow employing the synchronous approach allows reliable specification and implementation analysis. Evaluations apply to the target code in close accordance to the final implementations. Boolean equations are a simple target code for the control part. Measurement and analysis of C programs provide a database for reliable evaluation of worst execution time analysis, available at early stages in the development process. The combination of OO design with the synchronous paradigm and timing analysis fills the mismatch of synchronous hypothesis and concrete system realization.

References

[1] Felice Balarin, Harry Hsieh, Attila Jurecska, Luciano Lavagno, and Alberto Sangiovanni-Vincentelli. Formal verification of embedded systems based on CFSM networks. In *Proc. 33rd Design Automation Conference*, 1996.

[2] A. Benveniste and G. Berry. The synchronous approach to reactive and real-time systems. *Proceedings of the IEEE*, 79(9), 1991.

[3] Grady Booch. *Object-Oriented Analysis and Design with Applications*. Benjamin Cummings, Redwood City, 2 edition, 1994.

[4] R. Budde. The Design and Programming Language synchronousEifel sE. Technical report, GMD-AiS, St. Augustin, 1997.

[5] Pai Chou, Ross B. Ortega, and Gaetano Borriello. The chinook hardware/software co-synthesis system. In *Proceedings of the Eight International Symposium on System Synthesis*, pages 22–27, Cannes, France, 1995.

[6] R. Ernst, J. Henkel, and T. Benner. Hardware/software cosynthesis for microcontrollers. *IEEE Design & Test of Computers*, 10(4):64–75, 1993.

[7] Daniel D. Gajsk, Nikil Dutt, Allen Wu, and Steve Lin. *High-Level Synthesis : Introduction to Chip and System Design*. Kluwer, 1992.

[8] Uwe Gläser and K. T. Cheng. Logic optimization by an improved sequential redundancy addition and removal technique. In *Proceedings IEEE Asian and South Pacific DAC*, pages 235–240, 1995.

[9] Rajesh K. Gupta and Giovanni de Micheli. A co-synthesis approach to embedded system design automation. *Design Automation for Embedded Systems*, 1(1-2):69–120, 1996.

[10] David Harel and A. Naamad. The STATEMATE semantics of statecharts. *ACMTSEM: ACM Transactions on Software Engineering and*, 5, 1996.

[11] Ivar Jacobson, Magnus Christerson, Parik Jonsson, and Gunnar Övergard. *Object-Oriented Software Engineering.* Addison-Wesley, 1992.

[12] A. Jerraya, M. Abid, and T. B.Ismail. Cosmos: A codesign approach for communicating systems. In *Third International Workshop on Hardware/Software Codesign,* pages 17–24, Grenoble, France, 1994. IEEE Computer Society Press.

[13] Sanjaya Kumar, James H. Aylor, Barry W. Johnson, and Wm. A. Wulf. *The Codesign of Embedded Systems: A Unified Hardware/Software Representation.* Kluwer, 1996.

[14] Y.-T. S. Li, S. Malik, and A. Wolfe. Performance estimation of embedded software with instruction cache modeling. In *International Conference on Computer Aided Design,* pages 380–387, Los Alamitos, Ca., USA, 1995. IEEE Computer Society Press.

[15] M. Chiodo, P. Giusto, H. Hsieh, A. Jurecska, L. Lavango, and A. Sangiovanni-Vincentelli. Synthesis of mixed software-hardware implementation from CFSM specifications. In *Proceeding of International Workshop on Hw-Sw Codesign,* 1993.

[16] J. Madsen, J. Grode, P. V. Knudsen, M. E. Petersen, and A. Haxthausen. Lycos: the lyngby co-synthesis system. *Design Automation of Embedded Systems,* 2(2), 1997.

[17] J. L. Peterson. *Petri Net Theory and the Modelling of Systems.* Prentice-Hall, Englewoods Cliffs, New Jersey, 1981.

[18] Axel Poigne, Matthew Morley, Olivier Maffeis, Leszek Holenderski, and Reinhard Budde. The synchronous approach to designing reactive systems. *Formal Methods in System Design,* 12(2):163–188, 1998.

[19] Paul Stravers. *Embedded System Design.* PhD thesis, TU Delft, 1994.

[20] Kei Suzuki and Alberto Sangiovanni-Vincentelli. Efficient software performance estimation methods for hardware/software codesign. In *33rd Design Automation Conference,* pages 605–610, New York, 1996. Association for Computing Machinery.

[21] Steven Vercauteren, Bill Lin, and Hugo De Man. A strategy for real-time kernel support in application-specific HW/SW embedded architectures. In *33rd Design Automation Conference,* pages 678–683, New York, 1996. Association for Computing Machinery.

Co-simulation

Heterogeneous System-Level Cosimulation with SDL and Matlab

Per Bjuréus*
Axel Jantsch**
*CelsiusTech Electronics, Sweden
**Royal Institute of Technology, Sweden

Key words: Cosimulation, SDL, Matlab, System-Level Design, Heterogeneity

Abstract: Many systems consist of a signal processing and a control dominated part. The interaction of the data processing functions and a large variety of system-level control functions are often complex and with far reaching consequences. Thus, an early analysis and assessment of this interaction in a system level model is desirable. We propose a heterogeneous cosimulation environment with Matlab for the signal processing parts and SDL for the control-dominated parts. We describe a communication and synchronisation technique that allows the natural usage of Matlab vectors which often represent data samples over time periods, rather than single events at time instances. This makes the technique both natural to use and efficient in the simulation. We describe two modes of synchronisation, head synchronisation and tail synchronisation, and the conditions under which they can be used together.

1.   INTRODUCTION

In the specification and design of dedicated digital signal processing (DSP) systems, the data processing and the system level control parts have traditionally been separated. This was justified because the main problems were in the functional complexity and the tight timing constraints of the data processing, while system level control were comparably simple and were typically added later during the implementation phase. Tools like Matlab, SPW, and COSSAP aid the designer in developing the signal processing al-

145

J. Mermet (ed.), Electronic Chips & Systems Design Languages, 145–157.
© 2001 *Kluwer Academic Publishers.*

146

gorithms and refining them to bit true models. From there C or VHDL code is generated or written for a custom hardware or a DSP based software solution. At this level, the system control is added to the design.

Today the situation is changing, and requires a system level integration of control and data processing parts due to the following reasons:

1. While specification and implementation of DSP functions are still an important research area, it is a mature field. The integration of these functions in various configurations and with other complex control dominated functions becomes increasingly a major challenge.

2. Many products and product areas develop very fast, and it is difficult to predict the required standards and interfaces for a product when it is in the market one year after the development started. Therefore, products need a high level of flexibility and reconfigurability to adapt to many potential future situations in which the product will be used. The consequence is a complex control accommodating all different standards and variants.

3. Today technology allows us to integrate much more than the essential core functionality in a product. Many functions to enhance the user's convenience, the flexibility of usage, maintainability and online testability, etc. are included. These functions contribute considerably to system level complexity and can constitute up to 90% of the system specification documents.

4. A system level model, which includes both the data processing and the system control, is desirable to assess the interaction of these two parts in an early design phase.

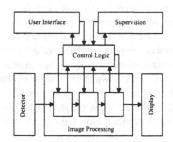

Figure 1. Image Processing System

Consider the image processing system for an infrared camera outlined in *Figure 1*. The system consists of an infrared detector, an image-processing unit, a user interface, a system supervision unit, control logic, and a display. The user interface allows the user to interact with the system and manually adjust the appearance of the image displayed. The supervision unit interfaces to external systems. It monitors detector temperature and external error conditions, controls lenses, and carries out self-tests. The control logic coordinates the system components. It reads and writes parameters to and from the image processing unit and handles user and supervision interaction. The detector scans the field of view, the image-processing unit transforms the image, and the display presents the image to the user. The image is updated several times per second (typically 25-50), which yields a data rate of several million pixels per second. The image processing is carried out in real-time and the throughput latency must be kept within certain limits, which implies strict real-time constraints. Meanwhile, the control logic is non-trivial, and may operate in several different modes, controlled by the user interface, the supervision unit, and feedback from the image processing system. It is crucial that the control logic behaves in a predictable way, and that it does not hang or enter undefined states that would impair the image processing.

When a system this complex is specified it is convenient to use different models of computation for the image processing part and the control logic. The image processing is effectively modelled using a dataflow paradigm whereas a finite state-machine model is better suited for the control logic. Since the control logic and image processing units are intimately connected to each other, the behaviour of the system is the combined behaviour of both units and their mutual interaction. The system modelling is often carried out by different design teams, each team specialised to work with one model of computation, e.g. dataflow or finite state-machines. It is therefore beneficial to provide heterogeneous system cosimulation that allows different design teams to simulate their contribution in an environment that considers the whole system. Simulation at an early design phase has several advantages. Errors are found quickly, and are easier and cheaper to correct. Feasibility and performance of different design solutions are easily explored, which results in a cheaper and more optimal design and shorter time to market.

We propose to use languages and tools, which are well established in their domains, namely Matlab [14] for the data processing parts and SDL [13] for the control dominated part. A major challenge is to integrate the timing and synchronisation concepts in a way that is intuitive to use. This is further complicated by our objective to take advantage of one of the essential ingredients of Matlab models, i.e. the usage of vectors and their transformations. Since vectors often represent data samples of a time period, their synchronisation with specific time instances and events in the SDL part has a subtle effect. We propose two synchronisation techniques, called head synchronisation and tail synchronisation, and show under which conditions they can be used together.

2.   RELATED WORK

Current approaches to system modelling can be divided into two groups, homogeneous and heterogeneous models. Homogeneous models are based on a single formalism or language such as VHDL, C++, CSP, SpecChart, etc. These languages are considerably rich and can typically be used far beyond their original scope. VHDL has been proposed as system-specification language [1], sometimes by extending it with advanced features such as communication facilities [15] or object-oriented concepts [4]. Similar attempts have been put forward for popular software languages such as C [3], C++ [2], Java [5][6], or SDL [7][8]. However, such homogeneous solutions come at a price. A language, which is well established in one community, is not always well received in another community. There are both accidental and essential reasons for this. The investment in a given language in terms of tools, competence, and existing designs is often so enormous, that an abrupt switch to another language cannot be justified. Also, the modelling concepts of general-purpose languages such as VHDL and C++ are not always a perfect match to the concepts of a given application problem.

For the two domains of control-dominated systems and signal processing, it is difficult to find a language that naturally accommodates both worlds. For these reasons heterogeneous frameworks have been proposed, those build on existing models and languages and devise techniques to integrate them. A very general and most influential framework is Ptolemy [9]. Ptolemy defines several models of computation such as discrete event or data flow domains. It provides a general mechanism for communication between different domains. A mechanism for communication and synchronisation between data flow and discrete event models has been implemented in Ptolemy, which transforms each single event on the border between the two domains. If we adopt this approach for the integration of SDL and Matlab, we would essentially lose the powerful vector handling in Matlab, which is both a user convenience and key to simulation efficiency. Hence, we chose to develop an alternative technique that avoids this disadvantage. Note, that our technique could be implemented in Ptolemy also, but for our particular purpose of integrating SDL and Matlab models, a more specific and less general solution is easier to realise.

Similarly, in CoWare [10] no mechanism is provided to communicate vectors in a synchronised way between different domains, which are described in C++, VHDL, and DFL (Data flow language). The communication is based on remote procedure calls, which can communicate any type of data, but the time attributes of this data have to be dealt with explicitly in the models by the designer.

VCI [11] is a cosimulation back plane system, which allows running several simulation engines concurrently. Marrec et al. [12] use it to provide an environment for VHDL and Matlab cosimulation in an untimed and a timed mode. The untimed mode does not utilise any timing information and is for functional validation only. The timed mode operates on a cycle-true timing model with concrete architectural components such as microprocessors. Our approach also allows an untimed functional simulation between SDL and Matlab. However, the timing model is more abstract than the timed mode in [12] because it is based on the timing behaviour of the input data, not on implementation components. Thus, it allows including the timing information of the input signals in the functional simulation, and based on this, facilitates the derivation of timing constraints for implementation components.

As a summary we can conclude that our proposal, in contrast to other approaches, addresses the problem of integrating the timing behaviour of input signals into a functional cosimulation of control dominated and data transformation parts modelled in SDL and Matlab, respectively.

3.   HETEROGENEOUS SYSTEM MODELING

Our target applications are typically embedded systems with a known interface towards their environment. The system is divided into a set of subsystems to make the system manageable with respect to size and complexity. A subsystem is modelled as a set of processes that operate concurrently and interact with each other. We consider two types of processes, control processes and dataflow processes. A control process interacts with its environment by exchange of control signals. Such a process has a state that changes over time as the process reacts to incoming control signals. A dataflow process is dominated by transformations of streams of data. A dataflow process consumes and produces data streams at a fixed rate. Streams flow from one dataflow process to another. The control processes and dataflow processes are allowed to interact with each other by exchange of control signals.

SDL employs an extended finite state-machine (EFSM) model of computation and is suitable for control systems modelling. The heterogeneous system specification is written using SDL at the top-level and to describe the structural hierarchy. SDL processes communicate with each other asynchronously through infinite FIFOs. The communication may or may not have a delay. In Matlab, data is transformed continuously from input stream to output stream. Streams are modelled as vectors or matrices, and transformations are modelled as functions with input and output parameters. All information exchange between the environment and the function must be passed as parameters.

3.1 Event Model

Processes communicate with each other by passing messages; this applies both to control and dataflow processes. A message may contain data or it may be empty. A message is associated with an event. An event consists of a message, the source and destination address of the message and a time stamp when the event occurs. The events are globally ordered in the system by the time stamp. An event that contains an empty message is referred to as a notification event. A dataflow message is referred to as a frame and the associated event is referred to as a continuous event. A control message is referred to as a signal and the associated event is referred to as a discrete event. Frames are used to represent streams and thus a frame always contains data and has duration. The frame data consists of a number of samples, which are the atomic elements of the stream. A stream is characterised by its sampling frequency, which specifies number of samples per time unit. Parameters can be passed to the dataflow model as control signals from the SDL model, and parameters can be passed back to the SDL model as status signals.

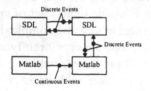

Figure 2. Event Model Outline

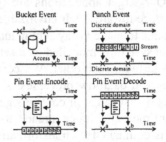

Figure 3. Event Types

Figure 2 shows the relation between process models and event types. The dataflow processes, specified in Matlab, communicate with other dataflow processes using continuous events whereas the control processes, written in SDL communicate both with control processes and dataflow processes using discrete events.

Discrete events are instantaneous and have no duration. The message of a discrete event is a signal that is allowed to carry a value. Discrete events are generated externally by the environment of the system and internally by the processes in the system. In SDL, timers are available that can be set for a timeout that causes a discrete event. Simulation time can be accessed during simulation and is used for synchronisation of the control and dataflow processes.

A stream often has an analogue source that can be measured by an electronic device. Examples of continuous streams are audio, video, microwaves, temperature etc. The system cosimulation targets embedded digital systems, and therefore only sampled streams are dealt with. A sampled stream is continuous since the samples in a stream are associated with a time and duration. At any time instant a unique sample is valid, i.e. the time and duration of the samples in a stream cover all time without overlapping. The continuously sampled stream is associated with a sample frequency, which specifies the duration of the samples in the stream. Different streams may have different sampling frequencies, but the sampling frequency of any one stream is constant.

An efficient way to represent streams is by using vectors whose elements correspond to samples. Using vectors allows large amounts of data to be processed efficiently. An individual sample is very similar to the discrete event presented earlier. However, an essential difference is that the sample has duration whereas the discrete event is instantaneous.

The Matlab models are assumed to operate on frames with fixed duration. A frame is provided to the Matlab model as a vector, where the order of the elements, i.e. the vector index, decides the time during which each sample is valid. A key concept when streams are converted to frames is stream splitting. When a stream is split, it becomes an ordered set of frames, where each frame is associated with a continuous event. Continuous events represent streams in the specification, but outside the Matlab environment they are modelled as SDL signals using discrete events.

When parameters are passed from the control model to the dataflow model and back as control and status signals, there are different ways to treat the signal depending on simulation timing.

Figure 3 depicts three different event types used to synchronise the message passing. The event types are referred to as bucket, pin, and punch events.

Bucket events are events that are collected into a "bucket". Each signal has its own slot and only the latest signal is stored, without a time stamp. When the bucket is accessed, the signal value is read and associated with an event that occurs at the time of the access. If the bucket is accessed and no event has occurred since the previous access, a default value is read.

Pin events are collected in a list. All events are recorded and marked with a time stamp. When the pin event list is accessed, it is translated between signals and a continuous stream vector with a predefined sampling rate. To encode a vector, signal values are "pinned" to elements in the vector and elements between "pins" are set to the value of the previous signal. To decode a stream vector with a known sampling rate, pin events are extracted and collected in an event list.

Finally, there is the punch event, which is used solely to "punch" out a value from a stream. The punch event immediately triggers a new discrete event containing the value of the stream at that particular time instant.

3.2 Synchronisation

A stream event is treated as a signal in the SDL simulator. The event is used to notify the dataflow models that data is available. The problem is to decide when the stream event should occur. First and maybe most naturally, one may cause the event when the first sample in the frame begins to occur. This will be referred to as head synchronisation, and can be viewed as information about the future. One may also do the opposite, and let the stream event occur when the last sample in the frame ceases to occur (which equals the time that the first element in the next frame begins to occur). This is referred to as tail synchronisation, and can be viewed as information about the past. "Begin to occur" and "cease to occur", relates to the fact that a sample has duration.

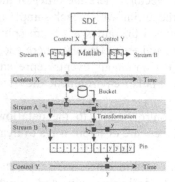

Figure 4. Head Synchronisation

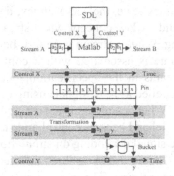

Figure 5. Tail Synchronisation

Figure 4 and *Figure 5* illustrate the head and tail synchronisation concept, which will be described in detail throughout the remainder of this section.

Head Synchronisation: When data is fed to a dataflow model, i.e. a Matlab function, the frame represents future data, and we expect the function to calculate new future data as its output, which propagates in the system. This means that signals that occur during the duration of the frame cannot be taken into account, since they have not yet occurred. This can be resolved by using the bucket event. During the duration of the frame, all input events are collected in a bucket, which is accessed with the next function call. The output from the dataflow model consists of frame and parameter vectors. The parameter vectors are translated to pin event lists, and the events can be transmitted to the rest of the system at the time they occur. *Figure 4* illustrates the head synchronisation technique. Filled squares represent the time of the events and a line connected to a stream event represents the duration of the frame. In the figure time extends from left to right. The first thing that happens is that the Matlab model receives frame a_1, and immediately transforms a_1 into b_1. The next thing that happens is that a control signal x is transmitted from the SDL model to the Matlab model. This signal cannot influence the transformation of a_1 to b_1 although it appears within the duration of a_1. Therefore, it is stored as a bucket event. Next, frame a_2 reaches the Matlab model, and the signal x can be accessed from the bucket and taken into account when transforming a_2 into b_2. The transformation of a_2 into b_2 yields a status signal y, which is returned in a vector that can be translated into a pin event list. The pin event list is used to transmit signal y from the Matlab model to the SDL model at the time it occurred.

Although head synchronisation does not allow input parameters to influence the calculation immediately, the stream samples can be accessed instantaneously with a punch event. The stream that is punched must be the input stream or the output stream of the dataflow model. Since the output stream already has been calculated, the punch event cannot alter the data.

Head synchronisation is most useful for dataflow models that lack input parameters and produces output parameters. A good example of such a process is data stream sources.

Tail Synchronisation: Tail synchronisation passes continuous events when they cease to occur. This means that input parameters can be collected in a pin event list, which can be translated to a vector and passed to the Matlab function. Output parameters cannot be sent when they occur, because when the function is called and the output parameter is calculated, that time has already passed. Thus, the output parameters are collected as bucket events. All output parameters are sent simultaneously after the function call.

Figure 5 shows the tail synchronisation technique. Filled squares still represent events, and lines connected to stream events represents frame duration, note however that the frame duration extends to the left of the event as opposed to head synchronisation. The first thing that happens is that signal x is transmitted from the SDL model to the Matlab model. The signal is collected in a pin event list. Next, frame a_1 reaches the Matlab model, the pin event list is translated to a vector that accompanies a_1 in the transformation, and the transformation of a_1 into b_1 is carried out. Finally, frame a_2 reaches the Matlab model and a_2 is transformed into b_2, which yields a status signal y that is collected in a bucket. The signal appears as a signal transmitted from the Matlab model to the SDL model at the time the transformation is carried out.

Since the input frames are passed to the dataflow model when they have ceased to occur, punch events are not allowed in tail synchronisation.

Tail synchronisation is most useful for dataflow models with input parameters that influence the data stream instantly. It is less useful for models that produce parameters.

Delay Requirements: It is possible to mix head and tail synchronisation in a system model, but certain rules must be obeyed when converting a stream from head to tail synchronisation and vice versa. The duration, or frame size, used by a dataflow process is denoted λt seconds. The computation delay, i.e. latency, of a dataflow process is denoted δt seconds. This means that a process collects data during λt seconds, and the delay from data input to output, or latency, is δt seconds.

Figure 6 shows two communicating dataflow processes where data flows from P1 to P2 via a signal route in SDL. P1 has duration λt_1, and P2 has duration λt_2. The computation delay of P2 is δt_2. Note that although the duration of a process requires that the input and output frame duration is equal, the duration of different processes operating on the same stream may differ. This situation is resolved using a buffer on each input port of dataflow processes, which provide means to concatenate or split incoming frames to match the process duration.

Table 1. Delay Requirements

Relation	P1	P2	Delay Req.
$\lambda t_1 = \lambda t_2$	Head	Head	None
	Head	Tail	None
	Tail	Head	$\delta t_2 \geq \lambda t_1$
	Tail	Tail	None
$\lambda t_1 < \lambda t_2$,	Head	Head	$\delta t_2 \geq (n-1) \lambda t_1$
$n\lambda t_1 = \lambda t_2$	Head	Tail	None
	Tail	Head	$\delta t_2 \geq n\lambda t_1$
	Tail	Tail	None
$\lambda t_1 > \lambda t_2$,	Head	Head	None
$\lambda t_1 = n\lambda t_2$	Head	Tail	None
	Tail	Head	$\delta t_2 \geq n\lambda t_2$
	Tail	Tail	$\delta t_2 \geq (n-1) \lambda t_2$

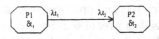

Figure 6. Process Communication

Table 1 shows the computation delay requirements depending on frame size for all combinations of synchronisation between two dataflow processes. If several streams are connected to the same dataflow process, all requirements for all combinations of input and output streams must be met.

4. IMPLEMENTATION

Wrappers are used to wrap up the dataflow components in the SDL specification in order to provide the synchronisation and data exchange during simulation.

Figure 7 depicts the relation between environments and wrappers used in simulation. The SDL model contains dataflow processes, each of which has a customised SDL wrapper. The SDL wrapper has access to a C-wrapper implemented as a set of C functions, which ultimately calls the Matlab engine through a set of C-library functions.

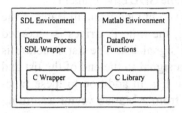

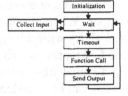

Figure 7. Wrappers and Environment   *Figure 8.* Wrapper Overview

Figure 8 shows a schematic overview of the SDL wrapper, which is implemented as a state-machine that handles input and output of streams and signals. The SDL wrapper declares a number of variables that are used throughout the signal processing. Two timers are declared: one that controls the dataflow function call, and one for output event synchronisation. Two variables are declared for each input stream signal, one holding the duration of the input frame, t, and one holding the ratio, r between the input stream frame duration and the duration of the process. The ratio is used to verify that all data is available before the function is called.

There is only one active Matlab engine, which executes all the dataflow processes. Vectors and variables, which are used only by Matlab functions, are never passed to SDL processes, which in turn allows for efficient simulation. Thus, typically only identifiers denoting the frames but not the frames themselves are passed between the SDL and Matlab models.

During initialisation, the Matlab engine is started, the duration D of the process and the sample rate of every stream, connected to the process, is determined. Those configuration parameters are defined in the Matlab environment. The frame ratio for each stream is derived from the sample rate of the stream and the duration of the model. After initialisation, the process enters the "Wait" state. Finally, the timer has to be set to give a timeout when the process is supposed to call the dataflow function for the first time.

This timeout should occur at time *now* for a process with head synchronisation, and at time $D+now$ for a process with tail synchronisation.

In the SDL wrapper there is a branch from all states for every input stream signal. When an input stream event occurs, a stream event counter is increased by the ratio between the input stream frame size and the duration of the process. This implies that a frame larger than the duration of the dataflow model is split into several frames, and frames smaller than the duration of the dataflow model are concatenated into larger ones. After the dataflow function has been called, the output streams are made available for receiving dataflow functions within the Matlab environment and a notification signal is issued for each output signal in the SDL environment.

For all input discrete event signals, there is a branch from all states in the SDL wrapper. When an input signal arrives, the time it arrives is recorded and stored along with the signal value in a signal list. The list structure is part of the C wrapper, which allows adding a signal to the end of the list and removing a signal from the beginning of the list. In addition to these functions, the list can be encoded to or decoded from a Matlab vector according to the pin event-handling scheme.

5. CONCLUSION

Assuming that a heterogeneous modelling style is beneficial for todays and tomorrows heterogeneous systems, we have proposed a cosimulation technique, which accommodates data flow and control intensive parts in a system, based on Matlab and SDL. The technique allows the natural and intuitive usage of Matlab vectors. The challenge of synchronisation between these two worlds has been addressed with two different synchronisation modes, head and tail synchronisation, which can be combined under certain conditions. Our model only considers timing imposed by environment constraints on streams and events, but not delays of implementation components. Thus, it is used for architecture independent modelling and facilitates the derivation of timing constraints for the implementation. An experiment that demonstrates the technique has been performed, the results were reported in [16]. It has been omitted here due to lack of space.

Future work will apply the technique to larger applications. Furthermore, a link to the implementation will be established by deriving timing constraints and estimating performance properties of implementation variants.

Acknowledgements: The Telelogic Tau tool suite was used for SDL entry and simulation, courtesy of Telelogic AB, Sweden.

6.   REFERENCES

[1] W. Ecker, "Using VHDL for HW/SW Co-Specification", pp. 500 - 505, European Design Automation Conference, September 1993.

[2] Bill Lin, "A System Design Methodology for Software/Hardware Co-Development of Telecommunication Network Applications", Proceedings of the Design Automation Conference, 1996.

[3] R. Ernst and J. Henkel, "Hardware-Software Codesign of Embedded Controllers Based on Hardware extraction", Proceedings of the International Workshop on Hardware-Software Co-Design, September 1992.

[4] Peter J. Ashenden, Philip A. Wilsey, and Dale E. Martin, "SUAVE: Extending VHDL to Improve Data Modeling Support", IEEE Design & Test of Computers, pp. 34-44, April-June 1998.

[5] Rachid Helaihel and Kunle Olukotum, "Java as a Specification for Hardware-Software Systems", Proceedings of the International Conference on Computer-Aided Design, 1997.

[6] James Shin Young, Josh MacDonald, Michael Shilamn, Abdallah Tabbara, Paul Hilflinger, and Richard Newton, "Design and Specification of Embedded Systems in Java Using Successive, Formal Refinement", Proceedings of the 35th Design Automation Conference, 1998.

[7] Jean-Marc Daveau, Gilberto Fernandes Marchioro, Carlos Alberto Valderrama, and Ahmed Amine Jerraya, "VHDL generation from SDL specifications", Proceedings of Computer Hardware Description Languages, April 1997.

[8] Bengt Svantesson, Shashi Kumar, Ahmed Hemani, "A Methodology and Algorithms for efficient interprocess communication synthesis from system description in SDL", Proceedings of the IEEE International Conference on VLSI Design, 1998.

[9] J. Buck, S. Ha, E. A. Lee, and D. G. Messerschmitt, "Ptolemy: A Framework for Simulating and Prototyping Heterogeneous Systems", International Journal of Computer Simulation, 1992.

[10] Ivo Bolsens, Hugo de Man, Bill Lin, Karl van Rompaey, Steven Vercauteren, and Diederik Verkest, "Hardware/Software Codesign of Digital Telecommunication Systems", Proceedings of the IEEE, vol. 85, no. 3, pp. 391 - 418, March 1997.

[11] C. Valderrama, A. Changuel, P. Raghavan, M. Abid, T. Ismail, and A. Jerraya, "A Unified Model for Cosimulation and Cosynthesis of Mixed Hardware/Software Systems", Proceedings of the European Design and Test Conference (ED&TC95), 1995.

[12] P. Le Marrec, C. A. Valderrama, F. Hessel, A. A. Jerraya, M. Attia, and O. Cayrol, "Hardware, Software and Mechanical Cosimulation for Automotive Applications", Proceedings of the Ninth International Workshop on Rapid System Prototyping, pp. 202 - 206, 1998.

[13] Jan Ellsberger, Dieter Hogrefe, and Amardeo Sarma, SDL - Formal Object Oriented Language for Communicating Systems, Prentice Hall Europe, 1997.

[14] MATLAB: High-performance Numeric Computation and Visualization Software. User's Guide, 1992.

[15] Petru Eles, K. Kuchcinski, Zebo Peng, and A. Doboli, "Hardware/software partitioning of VHDL system specifications", European Design Automation Conference (Euro-DAC), 1996.

[16] Per Bjuréus, Axel Jantsch, "Heterogeneous System-Level Cosimulation with SDL and Matlab", Proceedings of Forum on Design Languages (FDL), 1999

VHDL-based HW/SW Cosimulation of Microsystems

Vincent Moser, Alexis Boegli, Hans Peter Amann, Fausto Pellandini
Institute of Microtechnology, University of Neuchâtel, Rue A.-L. Breguet 2, CH-2000 Neuchâtel, Switzerland, http://www-imt.unine.ch, mailto:vincent.moser@imt.unine.ch

Abstract: This paper presents a pragmatic solution for the hardware-software cosimulation of microsystems. It allows a user to coverify a microsystem made up of custom digital electronics, processing cores and software. It can also be extended to the cosimulation of mixed-mode (analogue-digital) systems as well as mixed-nature (e.g., electrical-mechanical) systems. The custom hardware is described in VHDL. The software is developed in a dedicated environment and then translated into VHDL for simulation. The processor is described in VHDL as an Instruction Set Simulator (ISS) able to execute the software. A single tool, a VHDL simulator, is used to simulate the whole system. Its graphical user interface has been extended with a new software debug window. The approach has been demonstrated for Microchip's PIC16C5X microcontroller family.

1.   INTRODUCTION

Microsystems are a particular class of embedded systems which are small in size (up to some cubic centimetres) and which include components of various nature such as sensors, actuators, analogue and digital electronics, processors as well as software modules. Usually, the different parts are designed separately and put together during system integration. The lack of precise specifications and of communication between design teams often results in costly integration problems. Therefore, we claim that global simulation should be used during the whole design process [1]. The behaviour of the non-electrical parts (sensors, actuators) can be described using an analogue hardware description language like HDL-A, or as SPICE macro-models. The digital parts are described in VHDL and the whole system can be simulated in a mixed-mode EDA simulation environment. In order to avoid HW/SW integration problems, processors and software modules should also be taken into account for system-level simulation.

J. Mermet (ed.), Electronic Chips & Systems Design Languages, 159–168.
© 2001 *Kluwer Academic Publishers.*

We describe in this paper a pragmatic approach for HW/SW cosimulation which can be applied to global microsystem simulation. In section 2 we present various HW/SW cosimulation approaches and we explain the selection of one of them. In section 3, the proposed solution is presented with more details. The processor is described in VHDL as an instruction set simulator (ISS) while the software - developed in assembly language - is translated into a simple VHDL model of a ROM. The VHDL environment has been extended to offer software engineers basic debug facilities. Due to the integration in a VHDL context, the extension towards mixed analogue-digital simulation is straightforward. The approach is shown for Microchip's PIC16C5X microcontroller family in section 4 with the help of an example. The results and limitations and future work are briefly commented in section 5, followed by our conclusions.

2. HARDWARE-SOFTWARE COSIMULATION

Several HW/SW cosimulation approaches were proposed in the literature. We briefly summarize the characteristics of three of them which seem quite promising.

2.1 Cosimulation with a hardware simulator and a HDL language

The processor and the program memory are described as behavioural models and instantiated in a system description. The simulation is performed using a HW simulator only so that it can be realized using common EDA tools. The machine code compiled for the target processor is executed by the model of the processor which allows interactions between the program and the custom hardware to be observed easily. Moreover, if VHDL is used, the models of the processors and memories can be easily exchanged amongst simulators. However, the simulation can be quite slow and software debug facilities are usually missing.

This approach has been investigated by Calvez [2], Gupta [3], and Pelz [4]. The latter combines HW/SW cosimulation with mixed-mode simulation of electromechanical systems. In a variant called TOSCA, Antoniazzi [5] uses a virtual instruction set (VIS), each particular processor being just modelled by its interface. In this last solution, the code is no longer compiled for the actual target processor and the simulation cannot be very accurate any more.

2.2 Cosimulation with a hardware simulator and a high-level programming language

With this approach, hardware tasks are described, e.g., in VHDL while software tasks are described in a high-level language like C. VTT Electronics [6] developed such a solution around a model of a RTOS (real-time operating system). The software modules are called by the VHDL simulator through a foreign kernel interface (FKI). The workstation debugger (e.g. dbx) can be used to investigate the SW. The simulation can be quite fast but, as the software is compiled for the processor of the host workstation, its interaction with the hardware cannot be studied precisely. However, the tool overhead is also limited because the SW development environment of the host workstation can be used.

2.3 Cosimulation with a hardware simulator and a dedicated instruction set simulator

Here, a HDL simulator is coupled to a software debug environment via a specific tool called a cosimulation kernel. As both the hardware and the software are simulated and debugged in a dedicated environment, the user can benefit from the native HW and SW debug facilities of the tools. Additionally, the accuracy of the software simulation is good because it is compiled for the target processor. However, the communication between the hardware simulator and the software debugger can slow down the cosimulation. Therefore, various optimization levels are provided to get a good trade-off between accuracy and speed. Commercial solutions are proposed on the market notably by Mentor Graphics [7] and Viewlogic [8]. Unfortunately, the set of processors covered is currently limited to high-end microprocessors. In addition, since these solutions are tool-dependent, the models are not portable.

In microsystem design, we address low-power, small-size systems of moderate complexity and therefore we do not use high-end microprocessors but rather 8-bit microcontrollers. As the models of the low-power processors we use are not available in commercial tools and as the size of the software is moderate (usually, some kilobyte of assembly code), we decided to follow the first approach *cosimulation with a hardware simulator and a HDL language*. In order to meet the needs of software engineers, we added software debug facilities to a VHDL simulator in the form of a new window and associated simulation commands.

3.   PROPOSED SOLUTION

The proposed HW/SW cosimulation solution has been demonstrated with the example of a PIC16C5X microcontroller. A codesign flow has been defined, an ISS model of the PIC16C5X microcontroller family has been implemented, and a VHDL generator for the ROM models has been developed.

3.1   Codesign flow

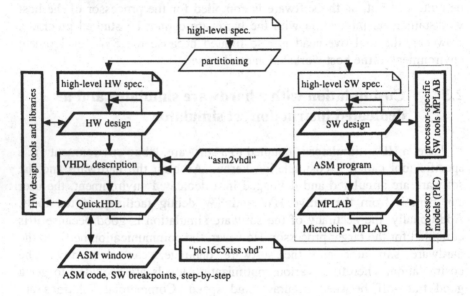

Figure 1. Codesign Flow

We propose the codesign flow of figure 1. It begins with the partitioning of the system specifications between hardware tasks and software tasks. Note that this top-level step is not addressed in this paper. The hardware tasks are implemented according to a usual VHDL flow. We work here with Mentor Graphics' QuickHDL. The software tasks are developed in a dedicated environment. For the PIC16C5X family, we use MPLAB by Microchip. When cosimulation is needed, the software has to be moved into the VHDL environment. Therefore, the machine code of the previously assembled and linked program is translated into VHDL. This is done by the tool *asm2vhdl* which produces a simplified asynchronous VHDL model of a ROM. Another VHDL model is necessary to execute this software. This model is the instruction set simulator *pic16c5xiss.vhd*. Finally, as the software cannot be properly observed and debugged in a native VHDL environment, we added a new software debug window *ASM window*.

The model of a complete microcontroller can be built easily. It is made up of a ROM model, an ISS model and some communication mechanism to the SW debug window. In the next paragraphs, we explain these particular models for the PIC16C5X family.

3.2   Modelling of the ROM

The model of the ROM is not available per se. It must be generated for each application specifically using the dedicated program *asm2vhdl*. The necessary information is all included in the listing provided by the PIC software development environment MPLAB. The model is behavioural and asynchronous. It stores the program in binary form but the original assembly source code is provided in comments as shown in table 1. For each 11-bit address read on the Addr port, a 12-bit instruction is written on the Data port.

Table 1. ROM VHDL code example

Assembly code

```
MOVF     mplr, W        ; Copy multiplier to W
BTFSC    mplr, 7        ; Test sign of multiplier
SUBWF    prodl, W       ; If negative, W <= (0x00 - W)
MOVWF    mplrcpy
```

MPLAB listing

```
000A 0208   00050   MOVF  mplr, W    ; Copy multiplier to W
000B 06E8   00051   BTFSC mplr, 7    ; Test sign of multip
000C 0093   00052   SUBWF prodl, W   ; If negative, W <= (0x00-W)
000D 0030   00053   MOVWF mplrcpy
```

VHDL code

```
elsif Addr = "00000001010" then
    OpCode := "001000001000" ;
            -- MOVF  mplr, W   ; Copy multiplier to W
elsif Addr = "00000001011" then
    OpCode := "011011101000" ;
            -- BTFSC mplr, 7   ; Test sign of multiplier
elsif Addr = "00000001100" then
    OpCode := "000010010011" ;
            -- SUBWF prodl, W  ; If negative, W <= (0x00-W)
elsif Addr = "00000001101" then
    OpCode := "000000110000" ;
            -- MOVWF mplrcpy
```

3.3   Modelling of the ISS

The behavioural model pic16c5xiss is an instruction set simulator for the PIC16C5X microcontroller family. It can fetch instructions from an external

ROM model, execute the instructions of the PIC16C5X instruction set and read or write the I/O ports.

The PIC16C5X 8-bit Harvard microcontrollers [9] have a clock pin, a reset pin, and I/O ports RA, RB, and optionally RC. RA is 4 bits wide while RB and RC are 8 bits wide. Internal registers can be divided into two groups: seven or eight Special Function Registers (SFR) and up to 73 general purpose registers. The program memory can contain up to 2K 12-bit wide instruction words. The RISC instruction set counts 33 12-bit wide instructions. A two-stage pipeline allows the microcontroller to fetch one instruction while executing the previous one. This way, all instructions are executed in one cycle, except for program branches.

In this model, the ROM is not included. Therefore, two additional ports have been introduced to communicate with the model of the ROM: the address port RomAddr and the data port RomData. An additional port LinAddr has also been introduced to communicate with the debug window for hardware-software cosimulation. The I/O ports, the registers and the pipeline effects have all been taken into account. Note that the clock signal of this instruction set simulator model is the instruction clock which is four times slower than the external clock of the actual PIC microcontroller. The PIC16C5X model we get is shown in figure 2 below.

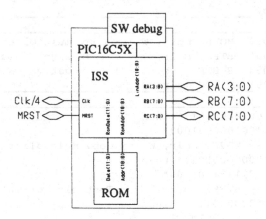

Figure 2. PIC16C5X model

3.4   Software debug window

The software debug window we developed is shown in figure 3. It allows users to view the assembly source code during the HW/SW cosimulation. They can set breakpoints and control the simulation by running either in step-by-step mode of for a given number of steps. The first column shows the line number; the second one marks the breakpoints; the third one has a

=> sign to point to the current line which is also emphasized in the main window. The debug commands are available either in the form of buttons or as commands which can be typed in the main QuickHDL window. Obviously, the simulation can also be driven from the HW side with the usual QuickHDL GUI. Note that in this SW view of the system, the time unit is the instruction cycle (step) while in the HW view physical time units are used. The correspondence between both domains is managed by the system.

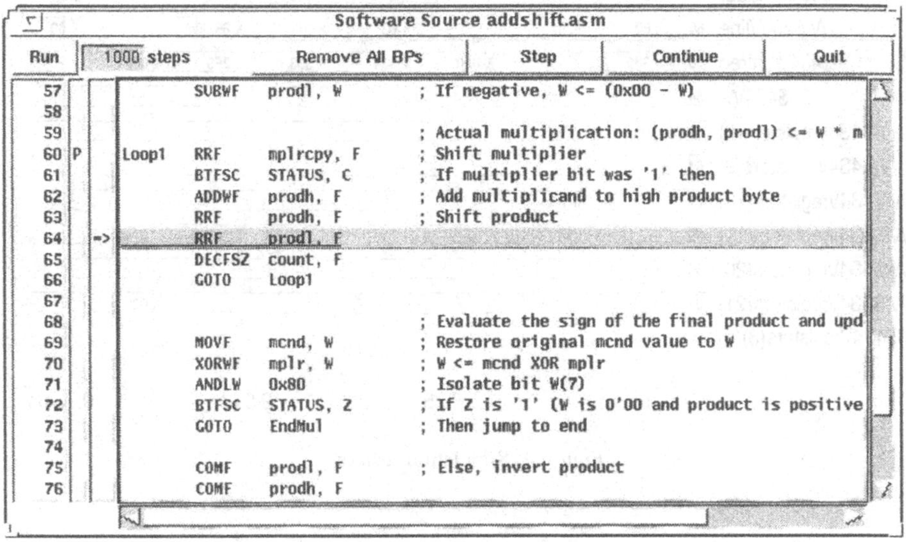

			Software Source addshift.asm			
Run	1000 steps		Remove All BPs	Step	Continue	Quit

```
57              SUBWF    prodl, W      ; If negative, W <= (0x00 - W)
58
59                                     ; Actual multiplication: (prodh, prodl) <= W * m
60 P   Loop1    RRF      mplrcpy, F    ; Shift multiplier
61              BTFSC    STATUS, C     ; If multiplier bit was '1' then
62              ADDWF    prodh, F      ; Add multiplicand to high product byte
63              RRF      prodh, F      ; Shift product
64 =>           RRF      prodl, F
65              DECFSZ   count, F
66              GOTO     Loop1
67
68                                     ; Evaluate the sign of the final product and upd
69              MOVF     mcnd, W       ; Restore original mcnd value to W
70              XORWF    mplr, W       ; W <= mcnd XOR mplr
71              ANDLW    0x80          ; Isolate bit W(7)
72              BTFSC    STATUS, Z     ; If Z is '1' (W is 0'00 and product is positive
73              GOTO     EndMul        ; Then jump to end
74
75              COMF     prodl, F      ; Else, invert product
76              COMF     prodh, F
```

Figure 3. SW debug window

4.   SIMULATION EXAMPLE

A complete PIC16C57 microcontroller has been modelled. In this example, an 8-bit "add-shift" multiplication algorithm has been implemented. The ROM model was generated from the corresponding assembly language program as an architecture named "addshift". Some simulation results are displayed in figure 4 below. First, some ports of the pic16c5xiss model are shown, including address and data ports. The contents of various registers like the register W are also visible. In particular, we see the two operands of the multiplication in the internal registers 8 and 9. The 16-bit wide result in contained in registers 20 and 19. Finally, register 21 is a loop counter used by the multiplication algorithm. The result is available when this counter decrements to zero. As expected, the result of the multiplication of 72 * 100 (16#48 * 16#64) is 7200 (16#1C20). To test this program the breakpoint facility and the step-by-step mode of the SW debug window were used.

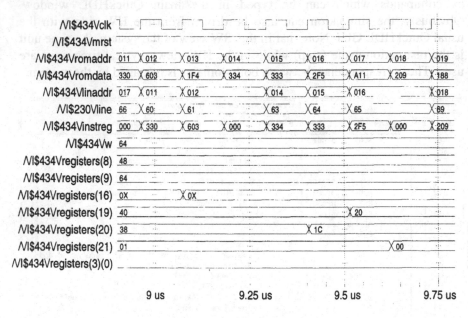

| | 9 us | 9.25 us | 9.5 us | 9.75 us |

Figure 4. Simulation results

5.   RESULTS AND LIMITATIONS

In the example above, the model of the PIC16C57 microcontroller was simulated without additional circuitry. We achieved a simulation speed of about 4000 instructions per second on a Sun Ultra 1/170. In another application, we simulated a mixed-mode, mixed-nature microsystem. It included a pressure sensor - modelled at behavioural level in HDL-A -, some analogue electronics - described in SPICE format -, and a PIC16C57 with a signal processing algorithm coded in assembly language. The simulation of the complete microsystem was done with Mentor Graphics Continuum-QuickHDL coupled with our cosimulation solution and we achieved 700 instructions per second on the same workstation.

Currently, the approach is limited to the simulation of assembly programs for the PIC16C5X microcontroller family. Several microcontrollers can be included in the system but only one of them can be "cosimulated" with the software debug window at a time. The remaining ones are simply made up of an ISS and a ROM model. When mixed-mode cosimulations were performed with Continuum-QuickHDL, we found out that some debug functions did not work properly because this tool does not behave exactly like the native QuickHDL.

The foreseen extensions include the development of ISS and ROM generators for other microprocessors as well as the development of a solution for ASM-C software integration.

6.   CONCLUSION

It has been shown in this paper that an acceptable HW/SW cosimulation solution for small to medium complexity systems, like microsystems, can be implemented in a VHDL framework. Even if the software is supposed to be developed in a dedicated environment, basic debug functions are offered. The advantages of the method are the limited tool overhead and the easy coupling to mixed analogue-digital simulation. Furthermore, it is quite easy to write an instruction set simulator in VHDL for a new microprocessor.

This approach can also be very attractive if portability is an issue. The ISS and ROM models can be used in any VHDL design environment. The communication mechanism related to the software debug window, however, should be fitted to the new VHDL simulator.

However, for the development of pure digital embedded systems with high-end microprocessors, a more sophisticated approach—like the one described above under *cosimulation with a hardware simulator and a dedicated instruction set simulator*—should be preferred, when available.

7.   REFERENCES

[1] A. Boegli et al., "System-Level Simulation Using Mixed-Nature Models", *1st Europe-Asia Congress on Mechatronics,* Besançon (F), October 1-3, 1996, pp. 12-17.
[2] J.P. Calvez et al., "Uninterpreted Co-Simulation for Performance Evaluation of Hw/Sw Systems", *IEEE 4th International Workshop on Hardware/Software Co-Design (Codes/CASHE),* 1996.
[3] R.K. Gupta et al., "Synthesis and Simulation of digital Systems Containing Interactive Hardware and Software Components", *Proceedings of the 29th Design Automation Conference,* June 1992, pp. 225-230.
[4] G. Pelz et al., "Hardware/Software-Cosimulation for Mechatronic System Design", *Proceedings EURO-DAC'96 European Design Automation Conference with EURO-VHDL'96,* Geneva, Switzerland, September 16-20, 1996, pp. 246-251.
[5] Antoniazzi et al., "The Role of VHDL within the TOSCA Hardware/Software Codesign Framework", *Proceedings EURO-DAC'94 European Design Automation Conference with EURO-VHDL'94,* Grenoble, France, September 19-23, 1994, pp. 612-617.
[6] J.-P. Soininen, "Cosimulation of Real-Time Control Systems", *Proceedings EURO-DAC'95 European Design Automation Conference with EURO-VHDL'95,* Brighton, Great Britain, September 18-22, 1995, pp. 170-175.

[7] "Hardware/Software Codesign at Mentor Graphics", http://www.mentorg.com/

[8] "Viewlogic HW/SW Co-Design", http://www.viewlogic.com/

[9] "PIC16C5X DataSheet, High Performance 8-Bit CMOS EPROM/ROM-Based Microcontrollers", DS30015M, Microchip Technology Inc., 1996, also available at http://www.microchip.com/

Modeling Interrupts for HW/SW Co-Simulation based on a VHDL/C Coupling

Matthias Bauer, Wolfgang Ecker, Andreas Zinn
Infineon Technologies AG, Corporate Development
Otto-Hahn-Ring 6 - 81370 Munich/Germany
phone +49 89 636 45515/45334/51734 (fax 44950)
e-mail: [Matthias.Bauer|Wolfgang.Ecker|Andreas.Zinn]@infineon.com

ABSTRACT

Interaction between hardware and software is often the reason for integration errors when combining both parts. Consequently, Hardware/Software Co-Simulation is required to keep with an increasing demand for an early verification of correct interaction. Using a VHDL/C Coupling, our approach of a modern testbench environment provides the possibility for Co-Simulation. At the beginning interaction was done by executing read and write operations on the software side resulting in CPU bus operations performed by a bus functional model running on a hardware simulator. To reach a higher level of interaction we extended the bus functional model to a peripheral functional model by including an interrupt handler. Accordingly, the hardware is able to react on events from the outside while sending interrupt requests to the software. In consideration of this approach most parts of the final C-software can be tested together with the hardware during an early state of the design flow.

MOTIVATION

From early days, software running on a software interpreter[1] interacted with some hardware devices[2][1]. This interaction is source of integration errors, even though two different design methodologies meet here. Thus merging the design of hardware and software was the dream of system designers and integrators for a long time. Due to the lack of computation power required for simulating a system including CPU-model, exhaustive simulations were often neglected. For that reason, hardware and software were developed sequentially, i.e. the software was implemented, tested, and debugged on an already finished hardware prototype. Another possibility is to use a prototype environment based on FPGAs which supports standard processor integration with realtime conditions and processor emulation [2].

Several trends forced, but also enabled, the merging of the hardware and software design process, mainly the virtual integration of both. This includes:

---

1. The piece of hardware, which is necessary to execute SW (E.g. CPU, controller, DSP,...).
2. The term hardware excludes in this paper the piece of hardware, which interprets software but which of course is hardware.

169

J. Mermet (ed.), Electronic Chips & Systems Design Languages, 169-178.
© 2001 *Kluwer Academic Publishers.*

1. The advance in computer hardware and operating system techniques:

The increasing speed of workstations enables the simulations of thousands or even millions of clock cycles. According to that, robust inter process communication capabilities for networks heterogeneous in both, computers and operating systems, supports easy distribution of operating system processes [3]

2. The upcoming and permanently improved technique of hardware description languages and their tools:

The MIL, IEEE, and IEC VHDL standard and later on the IEEE standardization of VERILOG focussed the effort of both, language designers and tool makers, on two languages only. Nowadays improved code generation techniques and advances in simulator kernels including cycle based techniques have been made or are currently under construction. In conjunction with the industry-standard SWIFT, which might be replaced by an already existing VERILOG PLI and a VHDL PLI currently designed by an IEEE working group, these improvements enable making efficient behavior models for simulation purposes of the hardware and the software interpreter in either VHDL, VERILOG, or C.

3. The dramatic increase of hardware as well as software functional test effort:

Also known as "Murphy's"-law [4], the complexity of functional test cases, which are required to ensure a minimum quality of the hardware and software, increase much stronger than chip complexity. Besides advanced test and testbench techniques, this requires the use of pieces of software to test the hardware and vice versa. This can only be done if hardware and software design process are synchronized and virtual hardware/software integration based on Co-simulation is possible.

4. The increasing part of software in so called embedded system design:

We currently estimate that 80% of the design effort of embedded systems inside SIEMENS is made for software only. It is expected that the software part increases soon to 90% or even more.

5. The increasing chip complexity:

According to "Moore's" law, the chip complexity, measured in transistors on one chip, increases 4 times every 3 years. This enables the integration of one or more software interpreter on one chip[1], which moves software even closer to hardware. Close integration of hardware and software design techniques is needed to support that trend. Major point is here also the capability of virtual integration based on Co-simulation.

Currently, VHDL[2] and C play a dominant role in digital embedded system design: VHDL for the hardware part, and C for the software part. New research approaches

---

1. More extremely spoken, putting software interpreters as complex pre-designed modules on a chip is currently besides memory and FPGA fields the only possibility to fill up the available chip area.

2. Of course also VERILOG. Our implementation of the interrupt handler was done in VHDL only. For that reason, we focus on VHDL for the remaining paper.

in Hardware/Software Co-Design start from a common specification (which may be C or VHDL alone) and use a fully automatic partition. They still need some time to get used as a state of the art design methodology (as currently e.g. VHDL based RT-synthesis).

As already mentioned, nowadays main focus for real designs lies in the early verification of the correct interaction of hardware and software, which is called Hardware/Software Co-Simulation. Subsequent performance estimation are performed based on the Co-Simulation results. Simulation is much too slow to allow a complete test of the software functionality (e.g. a complete RTOS together with the application program) together with the virtual hardware, but all hardware software interactions can be tested and a high confidence in low level software routines (often called firmware or embedded software) can be reached.

In this paper, we describe an extension of a Co-Simulation technique based on hardware simulation and virtual software execution (i.e. software directly running on a software interpreter in hardware which may differ from the final target and not on a model of the software interpreter). Up-front, the two major approaches for Co-Simulation are described and compared. Finally, before the outlook, we show a small application example.

CO-SIMULATION APPROACHES

Two major approaches for Co-Simulation are used in current designs:

1. Both, hardware and software interpreter exist as an executable model in an HDL simulation environment. The software exists as object code in an additionally modeled memory, which in turn is a RAM, loaded at the beginning of simulation, or a ROM, configured statically.

2. The hardware is represented by a model running on a simulator, however the C-softwre is executed as an own operating system process separated from the simulator. It is not important whether they run on the same computer and/or platform. Interaction is done by execution of blocking read and write operations operating on the software side, which imply the execution of CPU bus operations performed by a bus functional model running on the simulator. C-program and VHDL simulator communicate via operating system inter process communication (IPC) methods. This technique is describe in more detail in the next section.

Several alternatives of the basic Hardware/Software Co-Simulation principles exist, e.g. COSYMA: Co-Simulation and Co-Synthesis system which supports a processor simulator for runtime analysis and verification [5]. Anothers possibility is Donald Thomas' model and method for Hardware/Software Co-Simulation. In this case, a specification is mapped on a processor and several ASICs. Software runs on the target interpreter and comunicates with a hardware simulator via UNIX-IPC [6]. The main difference is the model representation of both, the hardware and the software interpreter. Alternatives include the representation as prototype, as configured hardware of an emulator, as real time modeler, or several levels of model abstraction ranking from gate level, i.e. propagation delay accurate models, over RT-level, i.e.

clock accurate and register true models, to instruction set models, i.e. operation sequence accurate model.

Both principle alternatives have their strength, whereas the advantage of one approach is mostly the disadvantage of the other approach.

1. The software execution by a VHDL-model of the software interpreter is able to represent exact timing, can handle interrupt as well as DMA (i.e. can be triggered by the hardware), and can also execute assembly language code.

2. The VHDL/C based approach allows to use the original software debugger, has very high software execution speed (almost real time), and requires relatively less HDL modeling effort for the bus operations.

To overcome some problems of the VHDL/C based approach, we developed and implemented an advanced VHDL/C message scheme. Before describing the modifications, we present the basic VHDL/C Co-Simulation techniques next.

BASIC VHDL/C - CO-SIMULATION TECHNIQUES

A typical message flow is shown in the message sequence diagram (MSC) in Figure 1.

Figure 1: MSC of a read operation with simplified CPU bus waveform. Time advances from top to down.

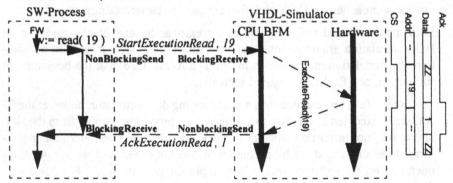

The semantic of the figure elements is shown in Figure 2. Messages are written in *italic* font and IPC based methods in **bold** font. On the right hand side, a simplified waveform of a bus-cycle is shown.

Figure 2: Semantic of message diagram elements

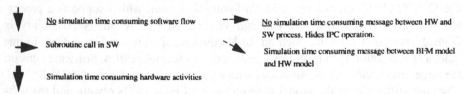

The interaction, also visualized as state diagram in Figure 3, works as follows:
First, both software and hardware simulation are started concurrently and do their task independently. The simulator evaluates the model until a reaction from the soft-

ware interpreter, represented by the bus functional model (BFM)[1] is required. In turn, software code is executed until access to addresses outside the program memory space, which resides on the computer running the software, is needed. This is the case, if for example an access to a physical register address of an ASIC must be performed. This access can not be executed directly due to the fact that the software does not run on its final target system. However, it is replaced by a read subroutine, or a write subroutine. This subroutine first sends a message via IPC to the BFM in the VHDL simulator.

The BFM waits blocking for such a message (thus prevents simulator from further evaluation) and continues execution (thus enabling continuation of simulation) after having received this message. In case, the simulator needs more time than the software to reach that interaction point, it can take the message directly without waiting. Next the message is decoded in the BFM and subroutines for generating bus protocol waveforms are called. These subroutines apply values to the computer bus modeled in VHDL, and in case of read, take data from the bus. After having finished the subroutines, the BFM sends an acknowledge (probably together with data) to the software via IPC and waits blocking for a message from the software again.

The simulator continues evaluation of the rest of the model. During the activation of the BFM, the software model waits blocking on an acknowledge from the BFM model. After having received the acknowledge, the read subroutine or write subroutine in the software process finalizes and probably returns a value. After that, software continues its execution until a new access to hardware resources is required.

It is important to note, that the software is always the master of the communication. First, software initiates communication and the simulator waits on an request. Synchronization is then completed by sending and waiting blocking on the acknowledge.

To make software quite a little more interactive to hardware, extended approaches including simulation time consuming NOP operations or iterative polling of hardware registers have been proposed. Polling may be combined with an iteration interval and a maximum iteration specification. Nevertheless, in spite of these improvements only some additional C-code can be tested but not the final C-software.

Advanced implementations map on the software side, calling subroutines for data access outside the program memory space to memory mapped IO. Here, the access to a specific address space is captured by the operating system, an operating system interrupt[2] is fired, and the corresponding interrupt routine manages data access with the VHDL simulator as described above.

---

1. A model representing the typical memory and external bus traffic only.
2. Please note, that this is an interrupt of the computer running the software and not an interrupt of the model.

174

Figure 3: State diagrams of HW/SW interaction (SW left + gray shaded, HW right, messages italic + dashed lines))

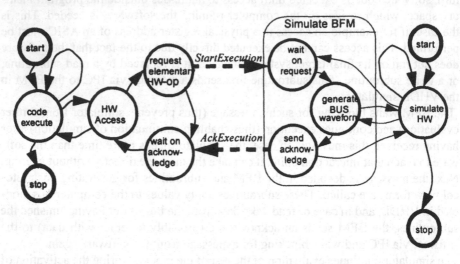

INTERRUPT HANDLING FOR VHDL/C CO-SIMULATION

For that reason, we extended the BFM of the CPU to a peripheral functional model (PFM) including interrupt handler and timer. Moreover we extended the message protocol for interrupt messages and included a message switch[1] into both, the C-softwre and the VHDL-PFM. The main idea of all implementation is that software execution can be interrupted either when a simulation time consuming operation (read, write, nop) is performed or the software is idle in the moment. However, it does not permanently perform nop operations, which would slow down simulation dramatically without having any benefit but which waits blocking for a message from the PFM in the simulator.

Figure 4: Block Structure of the PFM

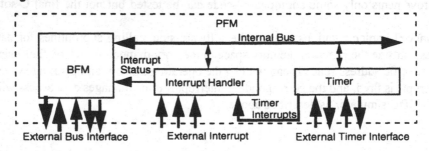

---

1. Now, software is no longer master of communication. The relationship changes dynamically depending on the kind of transmitted messages.

In detail:

1. The BFM is extended towards a PFM such that peripheral units are connected via an abstract modeled internal bus to the BFM. Since the peripherals are modeled as independent devices, concurrent actions of peripheral units and on the bus are supported. They react on both, messages from the internal bus and events from outside. The decoder of the BFM additionally differentiates between external addresses and internal addresses. In the first case, the bus protocol is generated, in the second case, an abstract message is sent to the peripheral model. Also, the PFM can initiate communication with software but only by sending an interrupt request instead of an acknowledge. The PFM can activate itself via PFM internal interrupt status signal. A block structure is shown in Figure 4 and an extended message flow for an interrupt occurring during a bus access can be found in Figure 5.

Figure 5: MSC of an interrupt handling

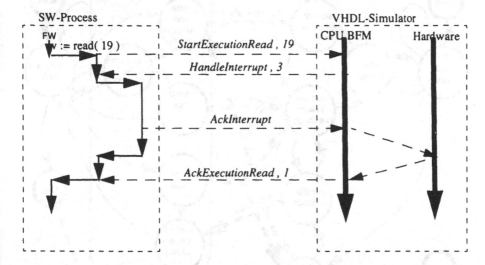

2. On the other side the software is only slightly modified. The subprograms handling the interface to addresses outside the software address space are embedded in a loop. After sending a message, the software waits blocking for an answer of the PFM. The only way to exit the loop is receiving the associated acknowledge to the sent message. Otherwise, the answer is interpreted as an interrupt message and one or a selection of interrupt routines is called. To allow several interrupt levels mixed with access to register of an ASIC or a peripheral during the execution of an interrupt, both interrupt handler and read/write subroutines can be called recursively.

3. A message switch in included in both, the BFM running on the simulator and the software. This message switch allows to change the master of the communication from the software side to the hardware side and vice versa.
   At the beginning, the software is communication master, i.e. can initialize communication as already mentioned before. This does not change, if a bus operation is requested and also if an interrupt handle is requested from the simulated

176

model.

However, when software moves to idle, the BFM running on the simulator becomes master. This changes immediately after another interrupt handle is requested from the simulated model. The software again becomes communication master. This changes again only, if the first interrupt routine entered after idle was left, i.e. the software returns to the idle state.

4. If software execution finalizes, it does not terminate. It waits in idle, i.e. calls a handler, which waits on an interrupt message from the hardware. This message is handled similar to the interrupt messages received instead of an acknowledge message as described above in point 2.

Figure 6: State diagrams of HW/SW interaction supporting interrupt handling (SW left + gray shaded, HW right, messages italic + dashed lines)

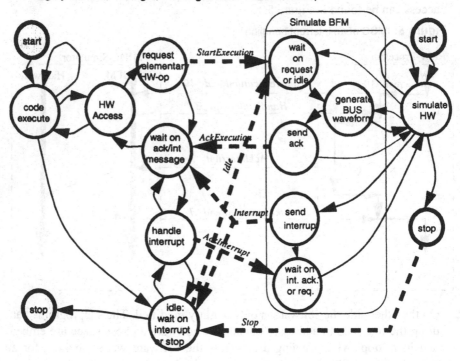

EXAMPLE

We implemented the interrupt model as described above exemplarily for an INTEL 386EX-Controller and used it first in small demonstration examples and afterwards for firmware Co-Simulation of a telecommunication system. For the reason of better understanding a small example should be discussed in more detail.

Using the modeled timer/counter unit we implemented a clock as described below. In this case the counter output is connected to one of the interrupt control unit's input. When the counter reaches one, the associated output is set and requests an interrupt service. Since configering the counter in a periodic mode, it is reloaded within the next rising edge of the internal clock and its output is forced to a low level again. The initial value of the counter is chosen that the resulting interrupt occurs every miilisecond. For this behaviour of the hardware both interrupt handler and timer/counter unit have to beconfigured as shown in Figure 7.

Figure 7: Initialisation of interrupt control unit and time/counter unit

```
void InitRealTimeClock()
{
    /* Initialize Interrupt Control Units */
    mpSocketPortWrite(ICU_ICW1_MASTER, "16#11#"); /* Edge triggered IR signals */
    mpSocketPortWrite(ICU_ICW2_MASTER, "16#20#"); /* Base interrupt vector is 32 */
    mpSocketPortWrite(ICU_ICW3_MASTER, "16#04#"); /* Slave ICU connected to IR2 */
    mpSocketPortWrite(ICU_ICW4_MASTER, "16#01#"); /* Fully nested mode, no AEOI */
    mpSocketPortWrite(ICU_ICW1_SLAVE,  "16#11#"); /* Edge triggered IR signals */
    mpSocketPortWrite(ICU_ICW2_SLAVE,  "16#28#"); /* Base interrupt vector is 40 */
    mpSocketPortWrite(ICU_ICW3_SLAVE,  "16#02#"); /* Slave ICU has ID 2 */
    mpSocketPortWrite(ICU_ICW4_SLAVE,  "16#01#"); /* Fully nested mode, no AEOI */

    /* Initialize Timer Counter Unit */
    mpSocketPortWrite( TCU_TMRCON, "16#35#" );
    mpSocketPortWrite( TCU_TMR0, "16#00#" );
    mpSocketPortWrite( TCU_TMR0, "16#00#" );

    /* Enable all interrupts */
    mpSocketPortWrite(ICU_OCW1_MASTER, "16#00#");
    mpSocketPortWrite(ICU_OCW1_SLAVE,  "16#00#");

    /* Set flag to indicate that the real time clock has been initialized */
    RealTimeClock_init = 1;
}
```

After detecting an incoming exception the interrupt handler sends an interrupt message to the software as described in Figure 6. In addition to this message, the corresponding interrupt vector number is transfered. On the software side the implemented switch descides whether the incoming message is an acknowledge to the previous command or an interrupt. Whereas a "MP_ACK" is returned in the first case, the firmware exception service routine in called in the second one as described in the listing of Figure 8.

Figure 8: Software Switch

```
char* WaitOnMpAck()
{
    char* pcBuffer;
    while ((pcBuffer = RecvSocketData(iGlobalSocket)) != NULL
            && strncmp(pcBuffer, MP_ACK, strlen(MP_ACK)) != 0)
        FW_Exception(pdBuffer);
    return pcBuffer;
}
```

Following, the interrupt service routine evaluates the interrupt vector number and calls the associated subroutine. In our case the real time clock has to be incremented as shown in Figure 8. After finishing the service routine the software sends an interrupt acknowledge to the hardware.

Figure 9: Interrupt Sevice Routine

```
void IntCounter0() /* increments the real time clock */
{
    if ( ++RealTimeClock_milliseconds >= 1000 ) {
        RealTimeClock_milliseconds -= 1000;
        if ( ++RealTimeClock_seconds >= 60 ) {
            RealTimeClock_seconds -= 60;
            if ( ++RealTimeClock_minutes >= 60 ) {
                RealTimeClock_minutes -= 60;
                if ( ++RealTimeClock_hours >= 24 )
```

```
           RealTimeClock_hours -= 24;
         }
       }
     }
   EndOfInterruptMaster();   /* end of interrupt */
 }
```

Conclusion and Outlook

We are currently extending the model towards peripheral communication units, e.g. a direct memory access handler and a serial interface unit. With the help of the implemented peripherals we are already able to handle complex applications. Furthermore, to become more flexible on the software side we are going to allow multi-tasking firmware routines.

As mentioned before, we implemented the PFM exemplarily for an INTEL 386EX-Controller and are now going to test our approach for another controller. After that, we plan to make a full controller model of the INTEL 386EX to relate the quite high coding effort for the peripherals.

Bibliography

[1] Steven E. Schulz, "Co-Verification Strategies In Hardware-Software Co-Design". Integrated System Design Magazine, August 1995.

[2] Genot Koch, Udo Kebshall, Wolfgang Rosenstiel, "A prototyping Environment for Hardware/Software Co-Design in the CORBA Project". Proceedings of the 3rd International Workshop HW/SW Codesign, Grenoble, France, September 1994.

[3] Steven E. Schulz, "Modelling Issues for Co-Verification". Integrated System Design Magazine, August 1995.

[4] Art de Geus, "Verification of Electronic Systems". Design Automation Conference DAC, Las Vegas NV, 1996.

[5] Rolf Ernst, Jorg Henkel, Thomas Bener, "Hardware-Software Cosynthesis for Microcontrollers". IEEE Design & Test for Computers, vol. 10, pp 64-75, 1973.

[6] Donald E. Thomas, J.K. Adams, H. Schmidt, "A Model and Methodology for Hardware-Software Codesign". IEEE Design & Test of Computers, pp 16-28, September 1993.

[7] Klaus Buchenrieder, "Hardware/Software Co-Design", IT Press Chicago, 1994.

SLD methodology

A COMPARISON OF SIX LANGUAGES FOR SYSTEM LEVEL DESCRIPTION OF TELECOM APPLICATIONS

Axel Jantsch[1], Shashi Kumar[2], Ingo Sander[1], Bengt Svantesson[1], Johnny Öberg[1], Ahmed Hemani[1], Peeter Ellervee[3], Mattias O'Nils[4]

[1] Royal Institute of Technology, Stockholm, Sweden
[2] Jönköping University, Jönköping, Sweden
[3] Tallinn Technical University, Tallinn, Estonia
[4] Mid Sweden University, Sundsvall, Sweden

Abstract: Based on a systematic evaluation method with a large number of criteria we compare six languages with respect to the suitability as a system specification and description language for telecom applications. The languages under evaluation are VHDL, C++, SDL, Haskell, Erlang, and ProGram. The evaluation method allows to give specific emphasis on particular aspects in a controlled way, which we use to make separate comparisons for pure software systems, pure hardware systems and mixed HW/SW systems.

1 INTRODUCTION

Language evaluation and comparison is difficult because of its large number of influencing factors, many of which are difficult to quantify. The outcome of most evaluations is therefore a subjective judgement which inherits its credibility from the individuals involved. Moreover, an educated debate about this judgement is rarely conclusive because of different priorities given by different people which are often not explicitly formulated and agreed upon. Thus, an argument by person X, stating that language A is superior to language B due to smaller synthesis results, would typically be countered by person Y by emphasising, that the simulator for language B is much faster allowing higher design efficiency. Although it is very difficult to quantify these and many other issues and to agree on defined priorities, more transparency and explicitness is absolutely necessary to make progress in the discipline of language and tool evaluation.

Based on a systematic method which is described in detail elsewhere [10], we present a comparison between several languages and illustrate, how giving high or low importance to a particular aspect affects the relative performance of a language.

J. Mermet (ed.), Electronic Chips & Systems Design Languages, 181–192.

2 EVALUATION METHOD

2.1 Scope of the Method

The evaluation method is targeted towards system specification languages for complex telecom applications. It is based on several assumptions:

- The design process defines separate phases for specification and design and requires separate specification and design documents. Pure requirements, functional or not, are also not considered part of the specification document. Hence, if requirements are explicitly formulated we assume that this is done in a separate requirements definition document.
- It is assumed that the specification document should capture the externally visible behaviour of the system and should avoid internal design and implementation decisions as much as possible.
- It is assumed to be an advantage if the specification document is amenable to analysis and synthesis tools. In fact, we assume that the more tools and methods can work with the document the better it is.
- As noted several times the target applications are complex systems, not simple systems that can be coded directly by one person in one week.
- The application area is telecommunication. We expect that complex electronic systems in other areas, e.g. in the automotive industry, exhibit similar characteristics, but we have not analysed other areas.

In the following two subsections we elaborate some of our assumptions concerning the purpose of the specification document and the application area.

2.2 Requirements for a Specification Document

In a product development process the specification is typically the first document, where the extensive discussion of many aspects of the problem leads to a first proposal of a system which shall solve the given problem. Hence, the purpose of the specification document is twofold:

1. It is a means to study if the proposed system will indeed be a solution to the posed problem with all its functional and non-functional requirements and constraints, i.e. to make sure to make the right system.
2. It defines the functionality and the constraints of the system for the following design and implementation phases.

From these two purposes we can derive several general requirements for a specification method.

A. **To support the specification process**: To write a specification is an iterative process. This process should be supported by a technique which allows the engineer to add, modify and remove the entities of his concern without a large impact on the rest of the specification.

B. **Analysable**: The specification should be analysable in various ways, e.g. by simulation, formal verification, performance analysis, etc.

C. **High abstraction**: The modelling concepts must be at a high enough abstraction level. The system engineer should not be bothered with modelling details, which are not relevant at this stage.

D. **Implementation independent**: The system specification should not bias the design and implementation in undesirable ways. System architects must be given as much freedom as possible to evaluate different architecture and implementation alternatives. Products are frequently developed in several versions with different performance and cost characteristics. Ideally the same functional specification should be used for all versions.

E. **Base for implementation**: The specification should support a systematic technique to derive an efficient implementation. This is in direct conflict with the requirements C and D.

2.3 Application Characteristics

Systems in a particular domain exhibit many characteristics to a different extent. In fact, the difference between different domains is usually not the absence or presence of characteristics but the degree of importance of different characteristics.

Our evaluation is targeted towards digital telecommunication systems. Such systems consist of signal paths and a reactive control system. A signal path consists of dataflow functional blocks operating on streams of data with potential high data rates. The reactive control system has typically lower performance constraints, is control dominated and has sometimes large memories for configuration data and system state. The following characteristics are important for this application domain:

- *Stream processing*: The system transforms streams of data according to simple protocol transformations or complex mathematical transformations with sometimes high performance requirements.
- *Complex control:* The system can be in many different states and modes, e.g. some of them are responsible for the normal operation, some for start-up and configuration, some for testing and diagnosis, others for detecting and handling error conditions, etc.
- *Well defined timing:* Environment and requirements establish defined timing constraints which are important for the control and the stream processing part.
- *Spatial distribution:* An integrated functionality is sometimes implemented at spatially separated locations.

- *Versatile interfaces:* The system is typically connected with the environment with various standardized interfaces and protocols.
- *Large memory:* The behaviour of both control and stream processing depends on the system's state and configuration, which sometimes require large memories used in an irregular manner.

Different parts of a telecom system exhibit different characteristics ranging from pure signal processing to control dominated system management.

2.4 Evaluation Criteria

The definition of evaluation criteria is as difficult and as important as the evaluation of languages itself, because the criteria and their relative weights basically determine the outcome of the evaluation process. We base our work on the studies described by Ardis et al. [1], Narayan and Gajski [2] and Davis [3]. To a limited extent we also used the criteria discussed by Nordström and Pettersson [4]. In all these reports a set of criteria is selected based on the assumption, that if a criterion is fulfilled by a language to a high degree, the language can be more effectively used and the design process will on average result in a better product than when the criterion is not fulfilled. This excludes the design process and the designer's skill from the language evaluation. It has the disadvantage that the dependence of the end product quality on a given criterion could be misjudged.

An alternative in the selection of criteria is taken by Lewerentz and Lindner [5]. There, the main criteria are properties of the resulting model, such as liveness and correctness. Although these are the criteria of ultimate importance, they are influenced by many factors related to the design process, which had to be identified and filtered out before establishing valid conclusions about the influence of the language on these properties.

Starting with the criteria discussed in [1, 2, 3, 4], we add new criteria, divide them into four groups, namely modelling, analysis, synthesis, and usability related aspects, as illustrated in figure 1. These groups are assessed independently from each other, which means a language is subject to four different assessments rather than one. The modelling group is further divided into aspects related to computation, communication, data and time. This division is based on the observation that these four modelling aspects can be analyses separately as discussed in [9].

This list of criteria is of course to some extent arbitrary, as is the case for any similar kind of evaluation. The list is perhaps not complete and the criteria are not orthogonal and independent from each other. Not all the criteria are on the same level, some could be merged into a single criterion. Others could be refined and split into criteria covering certain aspects in more detail.

We have introduced weights for each criterion to account for overlap

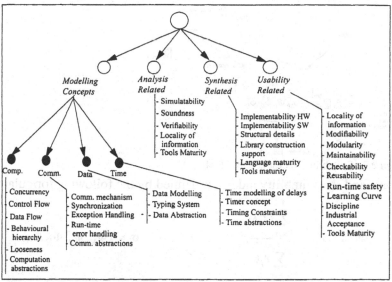

Figure 1. Evaluation criteria

between criteria and to emphasise the particular purpose of the evaluation, which is language evaluation as opposed to tool or design process evaluation. Furthermore, the criteria weights reflect the focus on specification rather than implementation. The weights allow to define priorities among the criteria but avoid implicit preferences by selecting or dropping certain criteria. The weight factors that we use in this comparison are listed in table 2. For more details on the criteria and their weights see [10].

2.5 Evaluation Mechanism

The objective of the evaluation is to get quantitative parameters to make it possible to compare the suitability of various languages for specification of systems in a given application domain. It is possible that the evaluation may conclude that language A is better than language B, language A is highly suitable for description but difficult to synthesize, or that language A is more suitable for large systems and language B is more suitable for small systems.

The method uses evaluation functions ϕ, which produce a suitability index depending on the evaluated language, the application area, the system size, and the design objectives, as illustrated in figure 2. Some factors are implicit in this scheme and therefore not explicitly visible in figure 2. The design phase affects the selection of criteria, the criteria weights, the context weights and perhaps even the language weights. For a different design phase the entire evaluation must be thought over again.

C, K, and L are numerical vectors with one element for each criterion. The

186

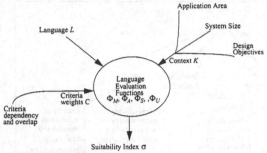

Figure 2. Suitability Vector is a four tuple

function Φ and the suitability index σ is defined by the following formulas:

$$\Phi(C, L, K) = \frac{\sum_{i=1}^{n} l_i c_i k_i}{\sum_{i=1}^{n} c_i k_i} \qquad\qquad \sigma = \langle \Phi_M, \Phi_A, \Phi_S, \Phi_U \rangle$$

By restricting the range to $K_i, C_i \in [0,1]$ and $L_i \in [-1, 1]$ the method guarantees a desirable metric as discussed in detail in [10]. To facilitate interpretation of these vectors and of the results the method uses following mapping of symbols to numbers:

K : (IRRELEVANT(IRR) $\leftrightarrow$ 0.0, UNIMPORTANT(UNI) $\leftrightarrow$ 0.25,
RELEVANT(REL) $\leftrightarrow$ 0.5, IMPORTANT(IMP) $\leftrightarrow$ 0.75, ESSENTIELL(ESS) $\leftrightarrow$ 1.0)

L : (VERY POOR (VEP) $\leftrightarrow$ –1.0, POOR $\leftrightarrow$ –0.5,
FAIR $\leftrightarrow$ 0.0, GOOD $\leftrightarrow$ 0.5, EXCELLENT (EXC) $\leftrightarrow$ 1.0)

Φ : (UNACCEPTABLE (UNA) $\leftrightarrow$ [–1, –0.5), UNSUITABLE (UNS) $\leftrightarrow$ [–0.5, 0),
SUITABLE (SUI) $\leftrightarrow$ [0, 0.5), PROPER (PRO) $\leftrightarrow$ [0.5, 1.0])

The purpose of these formulas is not to make the evaluation more objective but to make it more *transparent* by separating different influencing factors. The hope is that the identification and isolation of different influencing factors makes the assignment of proper weight factors easier. For instance it is easier to give a good answer to the question: "How important is the criterion of soundness for small sized control oriented applications during rapid prototyping?", than it is to answer: "How important is the criterion of soundness for my company or my department?". Thus, by splitting the large complex assessment into many smaller assessments the individual decisions become easier and the process of merging many small factors into one big decision is made more transparent and can be fine-tuned, rejected or accepted with confidence. It is needless to say, however, that the assignments of weights and the formula for computing the suitability index is still very subjective.

3 LANGUAGES UNDER EVALUATION

The languages under evaluation represent different paradigms. Erlang [6],

Table 1. Language assessment vectors

	L (Erlang)	L (C++)	L(Haskell)	L(VHDL)	L(SDL)	L(ProGram)
Structural hierarchy	GOOD	POOR	VEP	EXC	EXC	GOOD
Concurrency	EXC	VEP	FAIR	EXC	EXC	GOOD
Static processes	EXC	VEP	VEP	EXC	EXC	EXC
Dynamic processes	EXC	VEP	VEP	VEP	EXC	VEP
Control flow	EXC	EXC	EXC	GOOD	EXC	EXC
State machines	FAIR	FAIR	FAIR	GOOD	EXC	GOOD
Programming constructs	EXC	EXC	EXC	EXC	GOOD	POOR
Data flow	EXC	GOOD	EXC	EXC	POOR	POOR
Behavioural hierarchy	GOOD	GOOD	EXC	EXC	GOOD	GOOD
Looseness	GOOD	POOR	EXC	FAIR	GOOD	FAIR
Computation abstractions	GOOD	GOOD	EXC	GOOD	GOOD	VEP
Communication	FAIR	VEP	VEP	POOR	GOOD	GOOD
Synchronization	FAIR	VEP	VEP	GOOD	GOOD	GOOD
Exception handling	GOOD	FAIR	VEP	POOR	GOOD	GOOD
Run time error handling	EXC	FAIR	POOR	FAIR	FAIR	FAIR
Communication abstractions	POOR	VEP	POOR	POOR	EXC	FAIR
Data modelling	VEP	EXC	EXC	GOOD	GOOD	POOR
Typing system	VEP	FAIR	EXC	GOOD	GOOD	FAIR
Data abstractions	VEP	EXC	EXC	GOOD	GOOD	POOR
Timing modelling of delays	VEP	VEP	VEP	EXC	VEP	POOR
Timer concept	EXC	VEP	VEP	FAIR	GOOD	POOR
Timing constraints	POOR	VEP	VEP	POOR	POOR	POOR
Time abstraction	POOR	VEP	POOR	POOR	GOOD	POOR
Testability/Simulation	EXC	EXC	EXC	EXC	GOOD	POOR
Soundness	GOOD	VEP	EXC	POOR	GOOD	POOR
Verifiability	GOOD	FAIR	EXC	FAIR	GOOD	GOOD
Locality of information	GOOD	EXC	GOOD	GOOD	GOOD	POOR
Tools maturity- analysis	EXC	EXC	POOR	EXC	POOR	VEP
Implementability HW	VEP	POOR	VEP	EXC	VEP	EXC
Implementability SW	EXC	EXC	FAIR	FAIR	FAIR	GOOD
Structural details	VEP	VEP	VEP	EXC	VEP	FAIR
Library construction support	EXC	EXC	EXC	EXC	GOOD	VEP
Language maturity	GOOD	EXC	GOOD	EXC	GOOD	POOR
Tools maturity - synthesis	GOOD	EXC	POOR	EXC	FAIR	POOR
Locality of information	GOOD	EXC	GOOD	GOOD	GOOD	POOR
Modifiability	FAIR	GOOD	GOOD	GOOD	GOOD	GOOD
Modularity	GOOD	GOOD	GOOD	GOOD	GOOD	GOOD
Maintainability	GOOD	FAIR	GOOD	FAIR	FAIR	FAIR
Checkability	GOOD	POOR	GOOD	POOR	GOOD	EXC
Reusability	FAIR	GOOD	GOOD	GOOD	GOOD	GOOD
Run-time safety	GOOD	POOR	POOR	POOR	FAIR	POOR
Learning curve	GOOD	GOOD	FAIR	FAIR	GOOD	GOOD
Discipline	FAIR	FAIR	EXC	FAIR	GOOD	FAIR
Industrial acceptance	GOOD	EXC	VEP	EXC	GOOD	POOR
Tools maturity - usability	EXC	EXC	VEP	EXC	EXC	POOR

VHDL, SDL[12], and ProGram [13] have explicit concurrency; VHDL, C++, and SDL are imperative languages; Haskell [14], Erlang, and ProGram are declarative languages; C++ and SDL are object oriented languages; C++, SDL, Erlang, and Haskell have mostly been used for software development; VHDL and ProGram have been used for hardware development. One motivation for this selection was to cover different paradigms and aspects. Another practical reason was that these languages are well known by the authors.

In order to put the comparison on a solid foundation a realistic system has been modelled with all the languages, which contributes significantly to our confidence that the evaluation method is sound and the comparison is fair with respect to the given application domain.

The application example as supplied by Ericsson Telecom is an operation and maintenance system of an ATM network. ATM is an ITU-specified communication and switching technology for broadband services [7]. Operation and Maintenance (OAM) is part of ATM specifications that is responsible for detection of errors and performance degradation in the ATM network at switch level and to report it further [8]. A significant part of the OAM functionality in the ATM layer has been modeled in all languages [10]. The size of the models range from several hundred to a few thousand lines of code. However, the different OAM models cannot be compared with each other in a simple way due to several differences:

- The OAM models do not implement exactly the same functionality, even though they are very similar;
- The modelling style and the concepts used differ because the models have been developed by different persons with different objectives. This differences go far beyond what is induced by the use of different languages. For instance, the VHDL model uses bitvectors to represent ATM cells while the C++ model uses more abstract symbols; the Erlang model uses only static processes while the SDL model makes heavy use of dynamically created processes.
- The experience of the developers with the used languages varied.

For these reasons we do not attempt to compare the models in a superfluous way, like listing line numbers and development time. In a sense, the main result from the modelling activities is not the models but the analysis of language features with respect to a specific application domain. The application domain and the experience with the OAM functionality was always in the back of our minds when we analysed and discussed language concepts.

4 THE COMPARISON

We compare the languages in five different contexts. Table 1 shows the assessment vectors L for the languages. These are the results of the judgment

Table 2. Context vectors with different objectives

Criteria C		K(control SW)	K(mixed HW/SW)	K(pure functional)	K(pure HW)	K(simple HW)
Structural hierarchy	0.75	UNI	UNI	IRR	UNI	UNI
Concurrency	0.75	ESS	ESS	IRR	ESS	ESS
Static processes	0.5	ESS	IMP	IRR	IMP	IMP
Dynamic processes	0.5	ESS	UNI	IRR	IRR	IRR
Control flow	0.5	ESS	ESS	ESS	ESS	ESS
State machines	0.5	IMP	IMP	IMP	ESS	ESS
Programming constructs	0.5	ESS	IMP	IMP	IMP	IRR
Data flow	1.0	ESS	ESS	ESS	ESS	ESS
Behavioural hierarchy	1.0	ESS	ESS	ESS	ESS	UNI
Looseness	0.25	REL	ESS	ESS	ESS	UNI
Computation abstractions	1.0	ESS	ESS	ESS	ESS	IMP
Communication	0.75	ESS	ESS	IRR	ESS	IMP
Synchronization	0.5	ESS	ESS	IRR	ESS	ESS
Exception handling	0.75	ESS	ESS	UNI	ESS	IMP
Run time error handling	0.5	ESS	UNI	IRR	IRR	IRR
Communication abstractions	1.0	UNI	ESS	IRR	ESS	IMP
Data modelling	1.0	IRR	ESS	ESS	ESS	IMP
Typing system	1.0	REL	ESS	ESS	IMP	UNI
Data abstractions	1.0	IRR	ESS	ESS	ESS	UNI
Timing modelling of delays	0.75	IRR	IMP	IRR	IMP	IMP
Timer concept	0.75	ESS	IMP	IRR	IRR	IRR
Timing constraints	1.0	UNI	ESS	ESS	ESS	ESS
Time abstraction	1.0	UNI	ESS	ESS	ESS	UNI
Testability/Simulation	1.0	ESS	ESS	ESS	ESS	ESS
Soundness	1.0	IMP	ESS	ESS	ESS	IMP
Verifiability	0.75	ESS	ESS	ESS	ESS	ESS
Locality of information	1.0	ESS	ESS	ESS	ESS	UNI
Tools maturity- analysis	0.25	ESS	ESS	IRR	ESS	IMP
Implementability HW	1.0	IRR	IMP	IMP	IMP	IMP
Implementability SW	1.0	ESS	IMP	IMP	IRR	IRR
Structural details	0.5	IRR	IRR	IRR	IRR	IMP
Library construction support	1.0	ESS	ESS	ESS	ESS	ESS
Language maturity	0.5	ESS	ESS	IRR	ESS	ESS
Tools maturity - synthesis	0.25	ESS	ESS	IRR	ESS	ESS
Locality of information	1.0	ESS	ESS	ESS	ESS	IMP
Modifiability	1.0	ESS	ESS	ESS	ESS	IMP
Modularity	0.25	ESS	ESS	ESS	ESS	IMP
Maintainability	0.25	ESS	ESS	ESS	ESS	IMP
Checkability	0.5	ESS	ESS	ESS	ESS	IMP
Reusability	1.0	ESS	ESS	ESS	ESS	IMP
Run-time safety	1.0	ESS	UNI	IRR	IRR	IRR
Learning curve	0.25	REL	REL	IRR	REL	IMP
Discipline	1.0	REL	ESS	IRR	ESS	REL
Industrial acceptance	0.25	ESS	ESS	IRR	ESS	IMP
Tools maturity - usability	0.25	ESS	ESS	IRR	ESS	IMP

of one or several persons for each language. In particular there were 2 persons
to evaluate Erlang, 3 for C++, 2 for Haskell, 4 for VHDL, 2 for SDL and 2 for
ProGram. Table 2 shows the criteria vector *C* in the second column and the

context vectors K for the different objectives:

Control SW: Specification of large and complex control software as it is typical for the operation, control and management of telecom networks;

Mixed HW/SW: Specification of complex mixed HW/SW systems;

Pure functional: Specification of complex mixed HW/SW systems is similar to the "mixed HW/SW" context but with two distinguishing assumptions: (1) Explicit process level concurrency is irrelevant for the specification. One can argue that the partitioning into processes is in fact a design decision and should not be part of the specification [11]. (2) The focus is on research and factors such as tools maturity and industrial acceptance are considered to be irrelevant.

Pure HW: Specification of complex hardware systems;

Simple HW: Specification of smaller hardware systems. It is assumed that for smaller systems the need for high abstraction levels and constructs for complexity management is reduced. Thus, criteria such as behavioural hierarchy, looseness, and abstraction are deemphasized.

Tables 3 through 7 give the result of the comparison. Table 8 lists the languages which are suitable in each context, meaning, that the language performs SUITABLE or better in all four groups.

Table 3. Language comparison for large control SW

	L (Erlang)	L (C++)	L(Haskell)	L(VHDL)	L(SDL)	L(ProGram)
Φ_M	PRO	UNS	UNS	SUI	PRO	SUI
$\Phi_{MComputation}$	PRO	SUI	SUI	PRO	PRO	UNS
$\Phi_{MCommunication}$	SUI	UNA	UNA	UNS	SUI	SUI
Φ_{MData}	UNA	SUI	PRO	PRO	PRO	SUI
Φ_{MTime}	SUI	UNA	UNA	UNS	SUI	UNS
Φ_A	PRO	SUI	PRO	SUI	SUI	UNS
Φ_S	PRO	PRO	SUI	PRO	SUI	UNS;
Φ_U	SUI	SUI	SUI	SUI	SUI	SUI

Table 4. Language comparison for complex, mixed HW/SW

	L (Erlang)	L (C++)	L(Haskell)	L(VHDL)	L(SDL)	L(ProGram)
Φ_M	SUI	UNS	SUI	SUI	SUI	UNS
$\Phi_{MComputation}$	PRO	SUI	PRO	PRO	PRO	SUI
$\Phi_{MCommunication}$	SUI	UNA	UNA	UNS	PRO	SUI,
Φ_{MData}	UNA	PRO	PRO	PRO	PRO	UNS
Φ_{MTime}	UNS	UNA	UNA	UNS	UNS	UNS
Φ_A	PRO	SUI	PRO	SUI	SUI	UNS
Φ_S	SUI	PRO	SUI	PRO	SUI	UNS
Φ_U	SUI	SUI	SUI	SUI	SUI	SUI

At first it is surprising that C++ performs so poorly for control software applications. However, C++ neither supports concurrent processes nor any kind of timing, but both concepts are deemed to be important in this context. For C++ implementations typically the operating system provides these serv-

Table 5. Language comparison for complex, mixed HW/SW with low emphasis on concurrency

	L (Erlang)	L (C++)	L(Haskell)	L(VHDL)	L(SDL)	L(ProGram)
Φ_M	UNS	SUI	PRO	SUI	SUI	UNS
$\Phi_{MComputation}$	PRO	PRO	PRO	PRO	SUI	UNS
$\Phi_{MCommunication}$	PRO	SUI	UNA	UNS	PRO	PRO
Φ_{MData}	UNA	PRO	PRO	PRO	PRO	UNS
Φ_{MTime}	UNS	UNA	UNA	UNS	SUI	UNS
Φ_A	PRO	SUI	PRO	SUI	PRO	UNS
Φ_S	SUI	PRO	SUI	PRO	UNS	SUI
Φ_U	SUI	SUI	PRO	SUI	SUI	SUI

Table 6. Language comparison for complex pure HW systems

	L (Erlang)	L (C++)	L(Haskell)	L(VHDL)	L(SDL)	L(ProGram)
Φ_M	UNS	UNS	SUI	SUI	SUI	UNS
$\Phi_{MComputation}$	PRO	SUI	PRO	PRO	PRO	SUI
$\Phi_{MCommunication}$	UNS	UNA	UNA	UNS	PRO	SUI.
Φ_{MData}	UNA	PRO	PRO	PRO	PRO	UNS
Φ_{MTime}	UNA	UNA	UNA	UNS	UNS	UNS
Φ_A	PRO	SUI	PRO	SUI	SUI	UNS
Φ_S	SUI	PRO	SUI	PRO	SUI	UNS
Φ_U	SUI	SUI	SUI	SUI	PRO	SUI

Table 7. Language comparison for simple pure HW systems

	L (Erlang)	L (C++)	L(Haskell)	L(VHDL)	L(SDL)	L(ProGram)
Φ_M	SUI	UNS	UNS	SUI	SUI	UNS
$\Phi_{MComputation}$	PRO	SUI	SUI	PRO	PRO	SUI
$\Phi_{MCommunication}$	UNS	UNA	UNA	UNS	PRO	SUI
Φ_{MData}	UNA	PRO	PRO	PRO	PRO	UNS
Φ_{MTime}	UNA	UNA	UNA	UNS	UNA	UNS
Φ_A	PRO	SUI	PRO	SUI	SUI	UNS
Φ_S	SUI	SUI	SUI	PRO	UNS	UNS
Φ_U	SUI	SUI	SUI	SUI	PRO	SUI

ices. Hence, if one wants to evaluate C++ in combination with a particular operating system or if a particular aspect is not considered important, the vectors L and K have to be adjusted. In general the usage of particular tools or versions of a language will change the assessment, which makes apparent that it is difficult to draw general conclusions from specific evaluations. However, the evaluation method used here allows to analyse evaluation results and deviations from intuitive judgement.

Table 8. Comparison result

Context	suitable languages
Control software	Erlang, VHDL, SDL
mixed HW/SW	Erlang, Haskell, VHDL, SDL
pure functional	C++, Haskell, VHDL
pure HW	Haskell, VHDL, SDL
simple HW	Erlang, VHDL

5 CONCLUSION

We have presented a language comparison for specification of telecom systems. The main difficulty of such a task comes from the huge number of influencing factors and underlying assumptions and from the inherent subjectivity of the assessment by humans. We have not eliminated subjectivity and we cannot suggest a final conclusion but we have analysed different strengths and weaknesses of the languages and we have established causal relations between assumptions and evaluation results due to a systematic evaluation method. Each evaluation can still only be valid in a particular context and with respect to specific demands and objectives. However, we have shown a way to make an evaluation transparent and subject to detailed analysis and discussion by making all the assumptions and priorities as explicit as possible.

6 REFERENCES

[1] M. A. Ardis, J. A. Chaves, L. J. Jagadeesan, P. Mataga, C. Puchol, M. G. Staskauskas, J. Von Olnhausen, "A Framework for Evaluating Specification Methods for Reactive Systems - Experience Report", *IEEE Transactions on Software Engineering*, June 1996.

[2] Sanjiv Narayan and Daniel D Gajski, "Features Supporting System-Level Specification in HDLs", pp. 540 - 545, *European Design Automation Conference*, September 1993.

[3] Alan M. Davis, "A Comparison of Techniques for the Specification of External System behaviour", *Communications of the ACM*, pp. 1098 - 1115, September 1988.

[4] A.Nordström, H.Pettersson, *An Evaluation of Graphical HDL Tools with Aspects on Design Methodology and Reusability*, Ericsson, Sweden, Report JR/M-97:1676, 1997.

[5] Claus Lewerentz and Thomas Lindner, ed., *Case Study "Production Cell": A Comparative Study in Formal Software Development*, Forschungszentrum Informatik, Universität Karlsruhe, report no. FZI-Publication 1/94, Karlsruhe, Germany, 1994.

[6] J.Armstrong, R.Virding, M.Williams, *Concurrent Programming in Erlang*, Prentice Hall, 1993.

[7] M. De Prycker, *Asynchronous Transfer Mode solutions for broadband ISDN*, Series in Computer Communications and Networking, Ellis Horwood 1991.

[8] ITU-T Telecommunication Standardization sector of ITU Recommendation I.150, I.211, I.311, I.321, I.327, I.361, I.362, I.363, I.413, I.432, I.610.

[9] A. Jantsch, S. Kumar, A. Hemani, "The Rugby Model: A Framework for the Study of Modelling, Analysis, and Synthesis Concepts in Electronic Systems", *Proceedings of Design Automation and Test in Europe (DATE)*, 1999.

[10]A. Jantsch, S. Kumar, I. Sander, B. Svantesson, J. Öberg, and A. Hemani, *Evaluation of Languages for Specification of Telecom Systems*, report no. TRITA-ESD-1998-04, Department of Electronics, Royal Institute of Technology, Stockholm, Sweden, 1998.

[11]A. Jantsch and I. Sander, "On the Roles of Functions and Objects in System Specification", Proceedings of the International Workshop on Hardware/Software Codesign, 2000.

[12]A. Olsen, O Færgemand, B. Møller-Pedersen, R. Reed, and J.R.W Smith, *Systems Engineering with SDL-92*, North Holland, 1995.

[13]J. Öberg, *ProGram: A Grammar-Based Method for Specification and Hardware Synthesis of Communication Protocols*, PhD thesis, Dep. of Electronics, Royal Institute of Technology, TRITA-ESD-1999-03, 1999.

[14]J. Peterson and K. Hammond, editors, *Haskell Report 1.4*, http://haskell.org/.

HIGH LEVEL MODELLING IN SDL AND VHDL+

Francis Cook, Nathan Messer and Dr Andy Carpenter
MINT Group, Department of Computer Science, University of Manchester

Key words: VHDL+, SDL, Specification Languages, High Level Modelling

Abstract: This paper analyses the use of SDL and VHDL+ as high level functional
languages with regard to the levels at which these languages look at the
problem presented. The two languages are considered as candidates for
formalising the specification stage of the design process. The example for this
analysis is the design of the Universal Peripheral Controller, which involves
modelling the data flow between geographically separated processing nodes
and storage, across a controller (the UPC).

INTRODUCTION

The design process from RTL level behavioural descriptions through to
implementation is well defined and formalised, and hence mostly automated.
However, the design process above that is only just becoming formalised.
Initially at this level, a design language must capture the specification of the
system and the constraints within which it must operate. Ideally, the
specification will be executable so that it can be used to check the operation
of the system.

Subsequently, the design language must allow the partitioning of the
specification into executable blocks to allow the evaluation of architectural
possibilities before the detailed design begins. It should also allow the
specification and checking of the communication protocols that operate
between these executable blocks.

This paper looks at two languages, SDL [1] and VHDL+[2], that are used in
the early stages of design and assesses how well they fulfil the above

J. Mermet (ed.), Electronic Chips & Systems Design Languages, 193–203.
© 2001 *Kluwer Academic Publishers*.

protocol development and is in use in current design flows. VHDL+ was selected because it is the straw man for extensions to VHDL in the IEEE DASC study group on VHDL System and Interface based Design (PAR 1551). The work here is based on VHDL+ 4.0. The LRM for 5.0 became available after this paper was written.

LANGUAGE FEATURES

This section looks at the features of the two languages and their runtime environments. Their support of different design styles is examined in the next section.

Model Structure

In SDL, the structure of the model is tree-like set of container objects of various types. The top-most container type is **system**. This contains one or more **block** type objects. A block is, itself, a container for either **sub-block** or **process** type objects. Delaying or non-delaying **channels** connect blocks and sub-blocks together. Within a block, non-delaying **signalroutes** can connect sub-blocks and blocks containing processes together. Data can also be communicated by non-delaying variables. The definition of a process is either directly by its **behaviour** or by **services**. Services are sub-sections of a process that when combined together form a complete process.

Within the container tree, all of the objects contained by an object must be of the same type. For example, a particular block can contain sub-blocks or processes but not both. This tree forms a hierarchical name space. Hence, any type, variable or signal declared at a point on the tree is visible to all of the contained objects from this point down.

VHDL defines a model by the hierarchical composition of components. By separating the interface of each component, defined in an **entity**, from its operation/implementation, defined in an **architecture**, VHDL allows multiple descriptions of the details of a component. An architecture can both define behaviour and instantiate sub-components. Behaviour is defined by **processes** or concurrent statements, which are, in practice, shorthand notations for simple processes. Entities and processes are connected by **signals** or **shared variables**.

VHDL+ introduces a new design unit, **interface**, which besides allowing message passing permits the specification of the communications protocol used to connect entities together.

Both languages allow a hierarchical composition of a model. SDL provides a greater range of constructs and aligns the name space with this composition, whereas VHDL, and hence VHDL+, allows multiple descriptions of an object and so supports design refinement in a structured manner. SDL channels only define a path along which information can flow while signal connections (VHDL and VHDL+) and messages (VHDL+) have types, VHDL and VHDL+ is much more explicit about the way in which information flows between parts of the design. Both require directions to be spcified, SDL's channels and signal routes can be bi-directional.

Communications

Communication in SDL is by asynchronous signals that pass along channels and signalroutes. Although signals do not have a type, they can have parameters. These are used to transfer information. Signals have a point in time at which they are sent and received, at other times they do not exist. If a process is not in the appropriate state to consume a signal when it arrives, the signal will be lost.

It is possible to decompose channels into a top-level block containing other block and following the rules outlined in 2.1. It is not possible to decompose a signalroute.

Unlike SDL signals, SDL variables have a type and a value that is always available. Besides being visible within the section of the name space in which they are declared, variables can also be exported and imported between processes. Concurrency protection on variables means that simulations will always give deterministic results. Access to remote data objects can also be achieved via remote procedure calling.

VHDL signals have a type, a permanently accessible value, assignment delay and a history. The VHDL LRM also sets considerable semantics on signals so that when they have multiple pending events or are driven by multiple sources, simulations always give deterministic results. Variables in VHDL are either local and can not have concurrency problems, or are shared. It is up to the designer in VHDL to ensure that concurrent access to shared variables does not cause problems.

VHDL+ has an **interface** design unit that allows entities to communicate without connecting them via signals. Interfaces contain protocols that define top-level **messages** that are passed across the interface. Messages may be decomposed into lower level messages or at the bottom level linked to VHDL signals. In VHDL+ version 4.0, this decomposition is uni-directional, whereas in VHDL+ version 5.0, this decomposition can be bi-directional. This decomposition can include behaviour expressed in a finite state machine. Messages have a type and value and exist for the period of transmission, which can be a finite length of time rather than an instance.

The transmission of messages uses blocking **send** and **receive** statements. If an appropriate receive statement is not active when a message starts to arrive then it will be lost.

Both languages have normal programming language variables. However, the main communication is via signals and messages that are designed to provide deterministic communication between concurrent processes. Although there are some differences, SDL signals are closer in operation to VHDL+ messages than they are to VHDL signals. One of the main differences is that the correctness of a decomposition is not automatically checked.

Behaviour

In SDL, state machines capture the behaviour of a process. The receipt of a signal for which there is a waiting input triggers the transition between states. Process behaviour can be inherited with paths or entire sets of transitions overloaded to alter the state transitions.

In VHDL the behaviour of a process is defined by sequential statements. The progress through these statements can be interrupted until a signal is received, a complex condition is true, or a time period has expired. There is no way to create a second process whose behaviour is based on a previously defined process.

VHDL+ introduces higher level behavioural concepts of resources, choices and uncertainly in timing. A resource is an object that has a limit on the number of times it can be accessed concurrently. Once this limit is reached, subsequent access attempts are blocked until some of the current ones complete. A choice is a select statement where at runtime the simulator will randomly use one of the possibilities.

Resources are implemented in interfaces and **activities**. Within these, behaviour can be defined by infinitely nested blocks of sequential and parallel statements.

Despite the use of different description styles, the generally modelling capabilities of SDL and VHDL are similar. The provision of concepts used in the initial stages of design, e.g. resources and uncertainly, directly in VHDL+ is useful as it avoids the need for the designer to implement them themselves.

MODELLING ISSUES

This section examines the use of SDL and VHDL+ to model particular features that have occurred in the models that have been developed.

Elements of the SDL models developed are reproduced here, the VHDL+ models can be found in a previous paper [3].

SDL

Protocols

To illustrate this, we have tried to model the SCSI-II revision 10 protocol in SDL. Figure 1 represents part of the SCSI-II (revision 10) protocol, the entirety is too lengthy to reproduce here (interested parties should see [4]). The pieces of code shown deal with the devices when they compete for the bus.

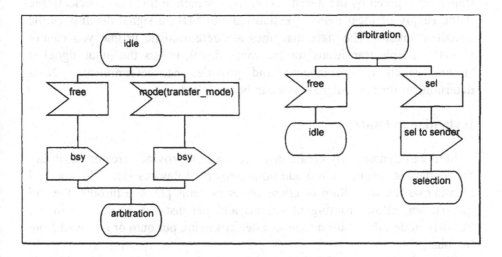

Figure 1. Part of SCSI-II Initialisation

There is no direct way to represent protocols, but the SDL signals would represent the protocol messages well. The protocol constitutes a series of actions that can be mapped to a state machine, which is how behaviour is modelled in SDL. The representation of the network protocols cannot capture the protocol at a low level i.e. the number of bits etc in the packet, so ASN.1 is used to do this. In situations where you model bus protocols, you have problems if you have to wait for bus lines to change status, requiring continuous monitoring. Receiving multiple signals at the same time is problematic in that while it is possible to receive many signals in a transition it is not easy to monitor the lines for a data value across the bus. If you are looking for a data value you require different methods of representing them, which means providing alternate signals or structs if you then have to use the same lines individually later on. While the continuous monitoring can be solved by using export variables, it detracts from the clarity and understanding of the model and may necessitate using a mixture of exporting all the bus lines from a process as variables and also using signals.

Control Flow Modelling

In SDL, this is very easily provided by the behaviour being defined as a finite state machine. Any action must take place after receipt of a signal. Figure 2 illustrates this. If a data signal is received then the local variable my_items is incremented, and an ack signal is sent. If an ack signal is received, a different transition takes place. Further control during a transition is given by the decision construct, which in this case checks to see if the supply of local items is exhausted. In SDL, a signal determines the transition taken. The state machines are deterministic in that you cannot choose multiple transitions for the same signal, unless the input signal is "none" which is spontaneous and provides non-determinism. Non-determinism for particular signals can be modelled.

Instantiation Issues

SDL can dynamically create instances, which provides greater flexibility for modelling resources and addition/removal of devices etc. This method also allows the modelling of client-server systems [1], and through the PId mechanism, allows routing of signals and per-instance handling. In the SCSI-II model, this is used to model devices being put onto or removed from the bus.

Figure 3 shows the SCSI-II model with the master process, which is modelling the bus, and the sda device that is instantiated. The numbers in parentheses refer to the numbers of instances present at the start of the simulation and the maximum allowed respectively, so (0,2) means no devices at the start with two devices maximum. The dashed arrow indicates which process instantiated which. The instances are created when a signal "device_add" is received by the "master" process along the "install" line.

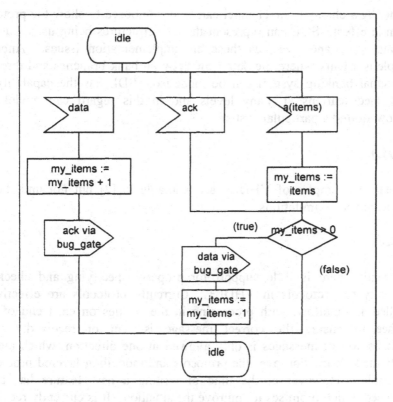

Figure 2. Control Flow

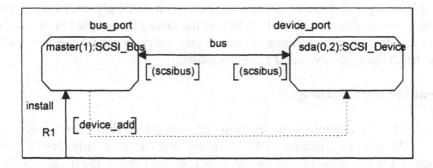

Figure 3. Instantiation

Dataflow Modelling

The way in which the actual data, that has to be transferred across a SCSI bus, can be modelled at the high level, is as a packet sent across a link. If we consider the application and the SCSI HD as two separate blocks linked together by a channel, the channel can be decomposed to show the protocol system in effect. SDL can represent the data structures being used, but not the data types and sizes, as these are implementation issues. Another example is a bank where the data flow between bank branches and through the internal banking system can be modelled. SDL has the capability to model specifications at many levels and in this regard can model the dataflow around a particular system.

VHDL+

Detailed examples of VHDL+ are available in [2] but not undertaken here due to space limitations.

Protocols

Currently there is little support for properly specifying and checking conformity to protocols in VHDL+. Currently protocols are effectively modelled as dataflow, with processing in the entities on each end of the interface to ensure the correct message is sent or received. The decomposition of messages is only allowed in one direction, which means only limited abstraction over the protocols and modelling layered protocols is possible. However at the time of writing a new feature has been announced which promises to improve the situation. It is currently required to specify which messages in the interface are "top level" messages, i.e. the components of the protocols.

Modelling full duplex protocols can be done with simply an interface and entities processing at each end. However modelling half-duplex protocols is trickier. In this circumstance, using another entity to model the bus itself and handle availability seems to be the solution.

Control Flow Modelling

VHDL+ is essentially an event-based imperative simulation language. An FSM can be constructed by ensuring that there is only one WAIT statement at the beginning of the behaviour, which handles the triggers of the FSM, and a switch statement which deals with working out the next state based on the current state, and another switch statement to work out the output based on the new state. Because VHDL (and VHDL+) were not designed as FSM languages, care must be taken. It is very easy to construct an FSM in which triggers can be missed (especially in an asynchronous FSM). It is also not as immediately obvious what is happening as in an SDL state machine.

Instantiation Issues

VHDL+ has no dynamic instancing abilities, so the number of each unit required must be known at compile time. VHDL+ offers the possibility of extending this soon. Messages and activities are considered a resource and currently if a message type of an interface, of an activity of an entity is in use, other attempts to use it are queued. However, we are promised concurrency on these resources that will allow the designer to specify how many concurrent calls may be in operation. We do not know exactly how this will work now, but it should offer the possibility of working out the number of units needed after rather than before execution.

Due to the lack of dynamic instancing in VHDL, an interesting problem was found with modelling the peripheral controllers. There are potentially over 2000 connections in operation over the controller. It seemed that creating this many instances of the control logic would be both costly in terms of simulation time, and would map most directly to a very inefficient implementation. Therefore, a design was conceived (Figure 4) whereby the control unit was a state machine, with no delays in the state machine itself, so it appears to run instantaneously. In fact, there are two such state machines, one for communications with the processing node, and one for communications with the peripherals. The time consuming tasks of communications, both with the outside world and the other half of the controller, are farmed out, and these are queued. The queuing is handled by putting these tasks in an activity, and triggering this, as activities provide the queuing. However, parameters for these activities had to be queued in an array, so the queuing is not completely built in.

Dataflow Modelling

This is an area that VHDL+ has made much easier to model simply and efficiently. Initially the flow of data and control messages around the system can be modelled with simple empty message types. As the model becomes more refined, parameters can be added to these messages in order to allow a more complex behaviour to be simulated.

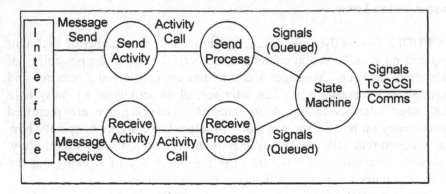

Figure 4. Full Duplex Communications

CONCLUSIONS

Both SDL and VHDL+ are discrete event simulation languages, with many similarities between their underlying semantics. However, it is the authors' view that SDL makes modelling at higher levels of abstraction easier than VHDL+. The link between an SDL model and it's implementation is less obvious that in VHDL+. This is an advantage for simply specifying a system, which is SDL's strength, but is part of why SDL does not fit into a standard design flow as easily as VHDL+. The flow from VHDL+ to VHDL is simple, and the refinement of abstractions, such as messages, through to low level details, such as signals, is easier to perform. This is both because the Supervise compiler produces VHDL, and because a designer can manually replace some high level abstractions with low level implementations, and still simulate the whole model together. The graphical nature of SDL also makes it a good language for initially specifying a system, as it aids the communication between a customer and a designer. One advantage VHDL+ has for specification is the ability to specify inexact timing and some non-deterministic behaviour.

The fact that VHDL+ is a superset of VHDL means that a designer can use the low-level VHDL constructs to capture behaviour that may be hard to capture in SDL's purely high level approach. One example of this is that everything must be seen as a state machine in SDL, whereas VHDL+'s imperative approach allows state machines, but is not tied to them. Another example is in modelling dataflow, where VHDL+ messages and SDL signals are similar. However, VHDL+ also inherits VHDL style signals that allow data to put on a signal and read when required by another entity, as well as modelling dataflow as a message send.

Although the capabilities of the two languages for control flow modelling are similar, specifying a state machine in VHDL+ is generally a slower and more error prone process than specifying one in SDL. One of SDL's big advantages is the ability to dynamically instantiate processes as required, whereas in VHDL+ it is necessary to statically instantiate everything at compile time.

Both languages have problems in abstracting over protocols. For SDL other notations exist to assist in this, such as ASN.1 and MSC. VHDL+ is still an evolving specification, and the new feature "transactions" that has been announce promises to improve this situation for VHDL+.

So in the authors' view VHDL+ significantly raises the abstraction level from VHDL, whilst making the flow from such a high level into a standard design flow easier. SDL on the other hand does not easily fit into the design flow, although for simply capturing the specification of a system it is an easier language to use.

ACKNOWLEDGEMENTS

The work described in the paper was funded by the EPSRC (grant number GR/L/28081) and supported by ICL High Performance Systems who have supplied us with design data, and both their SuperVISE and VISE systems.

REFERENCES

[1] A.Olsen, et. al., *Systems Engineering Using SDL-92*, North-Holland.
[2] ICL, *The VHDL+ Language Reference Manual Version 4.0*, 1999
[3] A. Carpenter, et. al., The Use of VHDL+ in the Specification Level Modelling of an Embedded System, *Proceedings FDL '98*.
[4] The SCSI-II (revision 10) Specification at http://www-micro.deis.unibo.it/~rambaldi/SCSI/SCSI2.html

ECL: A SPECIFICATION ENVIRONMENT FOR SYSTEM-LEVEL DESIGN

Gérard Berry
Ed Harcourt
Luciano Lavagno
Ellen Sentovich

Abstract   We propose a new specification environment for system-level design called ECL. It combines the Esterel and C languages to provide a more versatile means for specifying heterogeneous designs. It can be viewed as the addition to C of explicit constructs from Esterel for *concurrency* and *pre-emption*, and thus makes these operations easier to specify and more apparent. An ECL specification is compiled into a *reactive* part (an extended finite state machine representing most of the ECL program), and a pure data looping part. The first can be robustly estimated and synthesized to hardware or software, while the second is implemented in software as specified. ECL is a good candidate for specification of new behavior in system-level design tools such as Cadence's Cierto VCC tool[1]. ECL is especially targeted for specification of control protocols between data-computing behavioral blocks.

1.   OBJECTIVES

System-level designs are typically conceived as a set of communicating processes. The processes may communicate synchronously or asynchronously, may be control- or data-dominated, may have hard real-time constraints, and may be used in embedded systems. Such a wide variety of characteristics and requirements implies that there is no single language that can be efficient for specification. Nonetheless, it is desirable to be able to specify such designs in an integrated environment, so that the design as a whole can be both treated with a common semantics, at least at the communication level, and automatically synthesized, at least to the extent possible.

J. Mermet (ed.), Electronic Chips & Systems Design Languages, 205–212.
© 2001 *Kluwer Academic Publishers.*

For this reason, we propose the use of a new executable specification environment called ECL. The main idea is to combine two existing languages to create a specification medium that can benefit from the features of both languages and their existing well-developed compilers. In particular, we add the convenient and concise constructs from Esterel for concurrency and pre-emption to C.

A prototype ECL compiler has been completed and is currently being tested and further developed on some industrial examples.

2. BACKGROUND

2.1 ESTEREL

Esterel [5, 4] is a language and compiler with synchronous semantics. This means that an Esterel program has a global clock, and each module in the program reacts at each "tick" of the global clock. All modules react simultaneously and instantaneously, computing and emitting their outputs in "zero time", and then are quiescent until the next clock tick. This is classical finite state machine (FSM) behavior, but with a description that is distributed and implicit, making it very efficient. This underlying FSM behavior implies that the well-developed set of algorithms pertaining to FSMs can be applied to Esterel programs. Thus, one can perform property verification, implementation verification, and a battery of logic optimization algorithms.

The Esterel language provides special constructs that make the specification of complex control structures very natural. It is often referred to as a *reactive* language, since it is intended for control-dominated systems where continuous reaction to the environment is required. Communication is done by broadcasting signals, and a number of constructs are provided for manipulating these signals and supporting concurrency and signal pre-emption (e.g., parallel, abortion and suspension).

The Esterel compiler resolves the internal communication between modules, and creates a C program implementing the underlying FSM behavior. A sophisticated graphical source-level debugger is provided with the Esterel environment. While Esterel only provides a few simple data types, one can create and use any legal C data types; however, this is separate from the Esterel program, and must be defined separately by the designer. Pure C procedures and functions can be defined by the user and called from an Esterel program, but again there are definitions and code that must be written by hand by the designer.

2.2   C

The C language is ubiquitous. It is used as an application language (system-level programming), used for controlling hardware (e.g., drivers), and also used commonly as a hardware modeling language (e.g., instruction set simulators). However, it lacks control constructs that manage communication and concurrency between modules: one would have to implement these through hand-crafted data types and parameter passing. Nonetheless, C is a widely used language: the user-base is huge, there is much legacy code, and there are many robust compilers.

3.   ECL: ESTEREL + C LANGUAGES

3.1   OVERVIEW

ECL is primarily for *authoring new modules* in a system-level design tool. It is expected to be particularly useful for specification of of control-oriented, software-dominated glue communication functions such as protocol stacks. It supports a *mix of control (reactive) and data statements*, and automatically synthesizes the code needed for the interaction of these two. It compiles a *maximum subset* of the ECL specification into reactive modules in Esterel and subsequently into asynchronously communicating extended Finite State Machines called *Codesign Finite State Machines* (CFSMs [3]). CFSMs have a semantics that admits *robust optimization* and *synthesis to either hardware or software*, and their cost and performance can be estimated for a variety of possible subsequent implementations [3]. The rest of the program is compiled to data modules implemented in C and called by the Esterel modules.

3.2   SYNTAX

In terms of coding constructs and style, ECL simply combines Esterel and C. In particular, the concurrent and pre-emptive constructs of Esterel are added to C, along with communication by signals for controlling the flow.

The basic syntax of an ECL program is C-like, with the addition of the **module**. A module is like a subroutine, but may take special parameters called **signals**. The signals behave as signals in Esterel, and an equivalent subset of Esterel constructs are provided in ECL to manipulate them. As a simple example, the following code fragment:

```
#define data_size 80
typedef struct { int a, b } my_type;
module read_signal_data(input my_type IN_DATA, output int DONE)
{
```

```
int i, sum;
while (1) {
    for (sum = i = 0; i <= data_size; i++) {
        await(IN_DATA);
        sum += IN_DATA.a * IN_DATA.b;
    }
    emit (DONE, sum);
}
}
```

waits for **data_size** occurrences of the input signal, sums the two fields of each occurrence, and emits the DONE signal when all have been received. Note that in C, there is no natural construct for the communication through signals as done here; in Esterel, there is no automatic handling of the user-defined data type, and there is no explicit for loop. Though this can be easily specified with Esterel loops, for most designers it is more natural to use C looping constructs. In addition, Esterel loops must be reactive, that is, contain a halting statement. This could lead to an inefficient software implementation of loops that purely walk through, say, an array without waiting for external inputs, because it would require an FSM transition for each iteration that is more suited for hardware, rather than a straightforward loop-based software implementation.

3.3   SEMANTICS

The semantics of an ECL program are synchronous for each standalone, top-level single reactive module, just as with Esterel or any extended finite state machine language. At a higher-level, the modules are interconnected and will communicate via the semantics imposed at this higher level. In our current CFSM-based back-end implementation, a globally asynchronous semantics is applied at this network level.

The communication between parts of an ECL program, whether it be synchronous (within a top-level module) or asynchronous (between modules), is always done through signals (which may be valued). The decision about how to partition the design into synchronous individual modules communicating asynchronously is an implementation issue. We currently leave it to the designer to make such a choice, based on simulation and exploration at the specification level to aid in choosing the best implementation.

3.4   SUPPORT FOR PURE C AND ESTEREL

An ECL program typically is a mix of C and Esterel-like statements, but pure ANSI C and pure Esterel (with C-like syntax) are supported as subsets of ECL. This implies that legacy C code can be used in ECL-based system design.

The current compilation scheme for ECL translates as much of an ECL program as possible into Esterel, for full synthesis and optimization. In this way, we also maximize the subset of ECL that can be implemented as hardware, by being translated completely to Esterel first and CFSMs later. It is a subject for future work to explore schemes (more oriented towards legacy code handling and software implementation) in which only a minimal part of ECL, including only some reactive constructs (such as **abort**) is translated in Esterel, and the rest is left as C.

3.5   ECL COMPILATION

The prototype ECL compiler being implemented has two primary parts:

1. **Parser/Reactive Recognizer:** parses the ECL input into an internal data structure; traverses this data structure to recognize the reactive parts (Esterel-based statements), separate them, and write the result out in the form of C and Esterel code. This basically generates a description that can be used by the Esterel compiler to generate a top-level reactive FSM calling some (residual) C code. As previously stated, a maximum subset of the ECL program is compiled into Esterel, since it is this portion that can be estimated, aggressively optimized, and synthesized to hardware or software.

2. **Esterel compilation:** This part has two possibilities. The first is using the Esterel compiler to generate C code. Here, the code generated can be generic C code, or can be targeted for simulation using one of the Esterel simulators, or can be targeted for import to a commercial system design simulation tool. This flow has been tested with different simulators on different platforms and for a variety of test examples. The second possibility is using the Esterel compiler to generate CFSMs. This is a new part of the Esterel compiler being developed by the Esterel team in France and Cadence to compile Esterel programs to CFSMs and then to estimatable software and hardware. At a higher-level. the modules are connected via a globally asynchronous network communication scheme. As a side benefit. this compilation path uses the common

DC format, which implies that with it, there will be a connection to other synchronous languages besides Esterel. This second path is currently under development and test.

3.6   ON CONTROL AND DATA

One of the primary tasks of the ECL compiler is to recognize and separate the control and data parts. An ECL program will contain a mix of statements, manipulating signals and ordinary variables, communicating through signals, looping through computations, pre-empting operations, and calling external (C) functions. There are two types of loops: *reactive* loops which contain at least one Esterel await-type statement (e.g. **await (S);**), and *data* loops containing no such statements, and hence appearing to be instantaneous from a signal communication standpoint. A data loop is just like an ordinary loop in C, and is forbidden in Esterel since, with the notion of time, such a loop would be instantaneous (executing the same statements twice in the same clock tick). Data loops are allowed in ECL, but are compiled into separate C (inlined) functions called by the Esterel code.

The only ECL constructs that *cannot* be compiled into CFSMs are thus the *external (user-defined) functions*, and the *data loops*. In the treatment of complex data types, the ECL compiler generates the appropriate type definitions in Esterel and C, as well as the field and element access (inlined) functions called by Esterel and implemented in C.

The example above would be compiled into the following Esterel code:

```
type my_type;
function get_a (my_type): integer;
function get_b (my_type): integer;
procedure AUTOINCR (integer)();
module read_signal_data:
input IN_DATA:my_type;
output DONE:integer;
var sum, i: integer in
    loop
        sum := 0;
        i := 0;
        trap done in
            loop
                if i > 80 then exit done;
                await IN_DATA;
                sum := sum + get_a(IN_DATA) * get_b(IN_DATA);
                AUTOINCR(i)();
```

```
            end loop
          end trap;
          emit DONE (sum);
      end loop
  end var
  end module
```

A simple C header file would contain a set of macros implementing **AUTOINCR**, **get_a** and so on. In this case, there is no data loop, and hence the pure C code part is empty.

3.7   OPTIMIZATION AND SYNTHESIS TO HARDWARE AND SOFTWARE

A top-level single CFSM is synchronous and equivalent to an extended finite state machine. This implies that all the robust techniques for optimization (combinational and sequential logic optimization, optimization based on reachable states, etc), verification (both property verification and specification/design verification), and synthesis of FSMs can be applied. Furthermore, CFSMs can be synthesized indifferently to hardware or software. Thus, for the *reactive* parts of the ECL program, powerful techniques for implementation generation are available.

4.   CONCLUSIONS AND FURTHER DIRECTIONS

Since this paper was originally published, at FDL in 1998. considerable progress has been made. At that time, an ECL compiler prototype, proprietary to Cadence, was under test within their system-level design tools.

At the 1999 Design Automation Conference, a more extensive paper on ECL was published [6]. As of this latest writing, a new version of the compiler has been written in Java and is freely available on the web [2]. In addition, it has a smooth flow for integrating ECL models into the Cierto VCC system-level design tool by Cadence [1].

To summarize the capabilities. the ECL compiler compiles ECL programs into a maximally synthesizable subset; one important current direction for research is to synthesize only a *minimal* subset of the ECL program (the minimal reactive part), while leaving the rest in its C-code specification form. This style of compilation will be useful for importing legacy code, where the user would like to preserve the existing code as much as possible, while adding just enough "reactivity" to break this code into smaller pieces that interact through signals.

References

[1] For more information on Cadence's Cierto VCC product, visit /http://www.cadence.com/technology/hwsw/ciertovcc.

[2] The Java version of the ECL compiler has recently become available. Visit http://www.cadence.com/programs/na/research.shtml and follow the ECL project link.

[3] F. Balarin, M. Chiodo, P. Giusto, H. Hsieh, A. Jurecska, L. Lavagno, C. Passerone, A. Sangiovanni-Vincentelli, E. Sentovich, K. Suzuki, and B. Tabbara. *Hardware-Software Co-Design of Embedded Systems: The POLIS Approach*. Kluwer Academic Publishers, 1997.

[4] G. Berry. *The Foundations of Esterel*. 1998. To appear.

[5] G. Berry and G. Gonthier. The Esterel Synchronous Programming Language: Design Semantics, Implementation. *Science of Computer Programming*, 19(2):87–152, 1992.

[6] E. Sentovich and L. Luciano. ECL: A Specification Environment for System-Level Design. In 36nd *DAC*, pages 511–516, June 1999.

The MCSE Approach for System-Level Design

O. Pasquier, F. Muller, J.P. Calvez, D. Heller and E. Chenard
IRESTE, University of Nantes, France

Abstract   This paper presents the MCSE approach for the specification and design of embedded systems. The first motivation is to demonstrate its usefulness on the case study selected for the technical panel on the comparison of four languages for system level specifications at DATE 99. The MCSE solution for the crane control system is given first with the simulation results we have obtained. The second motivation is to stimulate more discussions on what engineers really need for system design. The term "language" can be understood as the interface between designers and tools or can be considered as a computational means inside tools. We show that we are in favor of a well-formalized notation directly usable and understandable by engineers able to describe solutions for verification first, then for performance studies and finally for product generation and/or synthesis. We also raise the ambiguity that exists on the possible levels of description when we use the term system-level specification and design.

1.   MOTIVATION

Embedded systems integrate high-complexity integrated components and deal with any kind of applications. Mastering on-going industrial applications needs the use of an appropriate system design and Hw/Sw CoDesign methodology to improve the quality of embedded systems and to reduce development cost and time. Such a methodology has to be supported by an appropriate language and tools to manage the projects, to capture the descriptions of solutions at various levels of abstraction, to verify these solutions, to produce the hardware and the software as automatically as possible, to validate resulting systems in their environments.

The mission of the SLDL committee is to support the creation and/or standardization of a language or representation, or set of languages or representations, that will enable engineers to describe single and multi-chip

213

J. Mermet (ed.), Electronic Chips & Systems Design Languages, 213–224.
© 2001 *Kluwer Academic Publishers.*

silicon-based embedded systems to any desired degree of detail and that will allow engineers to verify and/or validate those systems with appropriate tools and techniques. The languages or representations will address systems under design and their operating environment, including software, mixed-signal, optical, and micro-electronic mechanical characteristics [10]. The initiatives through several workshops on System Level Design [8], SLDL [10] and now SSDL lead to the necessity of considering case studies in order to have quantitative and qualitative criteria for comparing the different proposals. The panel held during DATE 99 concerning the comparison of four languages for system-level specification and design raises a very exciting challenge [6]. The portal crane system [5] has been considered as an appropriate example including most of problems engineers have to tackle nowadays.

We took the opportunity of this published case study to demonstrate our MCSE approach which is based on a published system specification and design methodology[1][9]. Recently this methodology has been investigated and enhanced in the context of the CoMES European project [4].

This paper is structured as follows. Section 2 briefly presents the MCSE methodology and Section 3 describes our solution to the crane system. Section 4 makes comments on the experiments presented at DATE 99 by four teams. We show in Section 5 that we are able to study significant properties of the system in order to produce its implementation. Section 6 raises the discussion on the objective of SLDL and conclusions are drawn in the last section.

2.   MCSE DESIGN METHODOLOGY

To help understand our system design approach, Figure 1 represents the MCSE design process. This is an ideal representation of the design flow as we do not voluntarily represent the verification activities and the backward flow to correct and/or enhance solutions. The flow is also limited to the front-end part of the complete development process.

System designers have to proceed according to a minimum of five steps: Requirement Definition, System Specification, Functional Design, Architectural Design, Prototyping. The design process must also enforce IP capitalisation and reuse at every design stage. Traceability in both forward and backward directions captures the relationships between requirements and all subsequent design data and helps managing requirement changes.

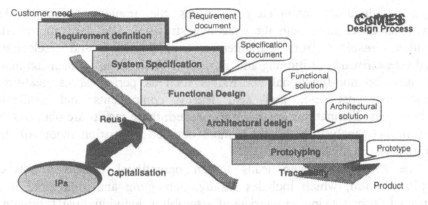

Figure 1. MCSE Design Process.

The *Requirements Definition* step is the process of understanding what the needs of all interested parties are and documenting these needs as written definitions and descriptions. The focus is on what problem the system has to solve, and so focussing on the world in which the system will operate, not on the system itself.

The purpose of the *System Specification* step is to express a purely external view of the system, (WHAT) starting from needs and user requirements. This step requires first a tight knowledge of the system's environment and then a complete modeling of the system behaviour. The functional specification can be based on data-flow diagrams (SA and SART methods), FSMs, StateCharts, SDL, etc., depending on the nature and the complexity of the problem. Specification models must be mostly executable to enforce their verification. This step appears usual but not so often used in companies.

The purpose of the *Functional Design* step is to find an appropriate internal architecture (which explains the HOW), but one according to an application-oriented viewpoint. The description based on a functional structure and the behavior of each function has to be technology-independent. Performances can be allocated to internal functions and relations between them.

The third step called the *Architectural Design* step aims at defining the detailed solution. It consists in searching, firstly for the executive support or hardware architecture and, secondly, for the organization of the software on each programmable processor. First, the functional description must be enhanced and detailed to take into account the technological constraints: geographic distribution (if necessary), physical and user interfaces. Timing constraints and performances are then analyzed to determine the hardware/software partitioning. The hardware part is specified by an executive structure or physical architecture. Mapping, which includes function allocation, completely describes the implementation of the functional description onto the executive structure. The most appropriate solution results from architecture exploration and selection, hardware/software partitioning and allocation. The criteria for decision must consider the non-functional constraints such as performances, real-time constraints, cost, etc; and also legacy components and available technologies. Estimations are therefore needed; these results are obtained by performance analysis based on a more detailed evaluation model of the solution(s).

The *Prototyping* step leads to an operational system prototype. Implementation, which includes testing, debugging and validation, is a bottom-up process since it consists of assembling individual parts, bringing out more and more abstract functionalities. Each level of the implementation is validated by checking for compliance with the specifications of the corresponding level in the top-down design process. Hardware and software implementations are developed simultaneously involving specialists in both domains, thus reducing the total implementation time. The complete solution can be generated and/or synthesized both for the hardware (ASICs and standard cores), for the software and for Hw/Sw interfaces. The resulting prototype is verified by co-simulation and emulation.

The two complementary and concurrent activities are: Capitalisation & Reuse, Requirement & Traceability Management. *Capitalisation and reuse* are essential activities for a correct and efficient exploitation of IPs. *Requirements traceability and management* refers to the ability to follow the life of a requirement in both forward and backward directions during the whole design process.

3. THE CRANE SOLUTION WITH MCSE

The case study includes functionalities and constraints which are representative of typical applications requiring embedded systems. It judiciously mixes time-discrete aspects with time-continuous behaviors as well as computations.

The specification document [5] describes the complete application and clearly identifies the system to be developed. The system environment is a portal crane (car and load) described by a fourth-order linear model (Fig 2). The embedded system includes sensors, actuators, two control strategies and diagnosis. Its objective is to properly move the load with the car to different desired positions.

By reading carefully the specification, we have easily defined the context diagram of the application. It consists of three parts:

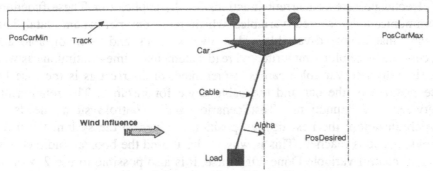

Figure 2. Representation of the problem.

- the crane part (CraneModel) which includes the car with its electrical command block and the load under the wind influence;
- the electronic system (ControlSystem) for monitoring, and controlling the crane movement;
- the test suite (TestScenario) which specifies the successive commands given to the crane system and models the environment conditions (wind influence) and break (or default) events.

These three concurrent parts are related by links representing the information given by the sensors and to the actuators. Figure 3 depicts the context diagram according to the graphical MCSE functional notation (please hide the internal models at this stage).

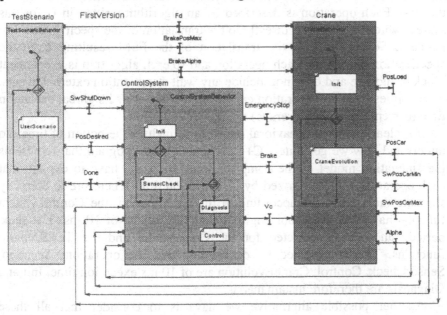

Figure 3. Functional structure of the solution.

Rectangles are concurrent functions for the application. These functions communicate by shared variables which represent permanent data. This means that a shared-variable value always exists and is modifiable and accessible by unblocking write and read statements. Time-continuous as well as time-discrete variables can be represented; of course this is the case for the position of the car and the cable angle for instance. The relationship between the functions TestScenario and ControlSystem needs a synchronization: the next desired position is given to the system when the previous one is reached. This is why we have used the boolean bidirectional shared control variable Done. Of course, it is also possible to use 2 boolean unidirectional variables Do and Done. This graphical functional model naturally includes the essential concepts of hierarchy, concurrency and communication.

From this description and to model the behavior of the system we need to refine the inside of each function in order to describe the complete specification. Again, by reading the specification document, we have defined the needed behavioral models which are depicted inside the 3 functions. TestScenario includes an Init operation followed by a loop on the Scenario operation. CraneModel is described by a similar behavioral model. ControlSystem includes the SensorCheck conditional loop after the Init operation followed by an infinite loop containing the Diagnosis and Control operations. These two latter operations are sequential for simplicity but can also be declared concurrent. The exit condition for SensorCheck is the end of this test. Each operation is described by an algorithm written in C for this system which corresponds directly to the translation of the specification. For instance, SensorCheck is the translation of the FSM resulting from the specification document. Each operation and so each algorithm is a statement block or module and does not include any wait statement on external events. This implies that each operation can be characterized by its global execution time (current, min, max, average).

It is clear that each behavioral model must include the notion of time. To understand how we generate a C++ executable code program that represents the simulation model of the complete application, we have to explain that each operation is characterized by its execution time. Therefore to correctly simulate the application according to the specification, the ControlSystem function has to evolve according to a sampling period of 10 ms. The same period can then be selected for the CraneModel and the TestScenario functions. This is obtained by considering that the operations Scenario, SensorCheck, Control, CraneEvolution are of 10 ms execution time. Init and Diagnosis are therefore instantaneous.

Another possible alternative we have is to consider that all these operation execution times are null, but the activation or invocation period of each function and so each loop is 10 ms. This is easily obtained with our method in considering that a function can be temporary according to a given activation period (here 10 ms). We have used this second method to produce our results. Properties such as execution times, activation periods, etc. are defined by what we call attributes attached to components of the model.

The design flow with the tools we have under development is the following one.

Figure 4. Design flow with the MCSE ToolBox.

The graphical editor helps capturing the functional and behavioral models of the solution. Figure 3 is the result printed by the graphical editing tool. The Attribute editor displays all attributes attached to each component. The designer can interactively modify values. The algorithm editor helps editing: the type definition of shared-variables; the local constants, types and variables in each function, the C algorithm of each operation. The C++ code generator produces an object-based model [7] for all graphical components of the functional model including their creation and execution methods. The execution method of a function is the translation of the graphical behavior model with the inclusion of the algorithms of its operations. The result is a multi-thread C++ executable program able to simulate the entire application including the test suite. The run of such a program needs a library of classes we have developed. Synchronizations and communications between threads are implemented by the WIN32 kernel of the Microsoft OS Windows 95 or NT.

Hereafter in Figure 5 you can see our results (PosCar, Alpha, VC) after the sensor check phase (more than 1900 s) and for the first test case.

The upper window depicts the position of the Car for a 100 s time frame. The influence of the wind is observable in the middle (Fd = 400 N). The 2 lower curves depict the evolution of the cable Angle and the motor command VC.

The main characteristics of our solution are the following ones:

- Number of C lines written by the designer: 530 lines;
- Number of the C++ program generated : 1860 lines (without libraries);
- Simulation time for the above 100s window: 51s;
- Design time of the solution from scratch: difficult to estimate because we were enhancing our tools at the same time. We think that less than 6 days is reasonable.

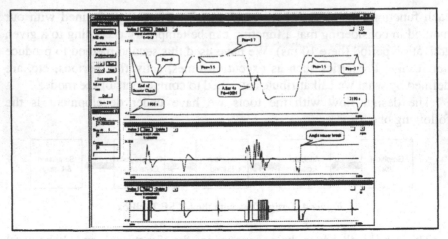

Figure 5. Results produced by our simulator and display analyzer.

4. OTHER EXPERIMENTS WITH FOUR STANDARD LANGUAGES AT DATE 99

This case study allows designers to compare different languages and approaches producing the requested results. Having attended the panel presentation and discussion at DATE 99, we can make a very partial comparison with the 4 methods experimented.

VHDL-AMS has appeared to us very appropriate to describe time-continuous, time-discrete behaviors together with finite-state behaviors. Simulation results were not provided; simulators were not available. Results were produced by the JAVA approach. The presentation showed that Java is an easy to use language to describe the whole application and to produce the graphical results. SpecC was also one experimental approach. No results were given. It seems that this is not due to a limitation of the tool which generates an executable C program, but more to a lack of time and also a lack of automatic control engineering culture for a computer science student. The approach by using the ADA language gives results. An event scheduler has been added to manage the execution of all tasks as well as to control the time evolution.

Our solution is an incremental technique based on graphical and textual "objects". The functional structure of the system can be easily modified; attributes are interactive parameters, and algorithms of operations are mostly not inter-dependent and can be quickly enhanced to take account of behavior evolution. The C++ program is automatically produced; this means that our tool produces the most difficult part of a software program which are its concurrent structure with communications and timed behaviors.

Even if we have found the example very interesting, we consider that it does not include all the problems usually encountered by designers. Most of problems include event-driven behaviors (reactive systems) and message-passing protocols over multi-point links. Task preemption and scheduling are also useful mechanisms for studying solutions before implementation.

5. MORE RESULTS ON THIS EXAMPLE

Starting from this example, one interesting question that designers can raise concerns what other results or properties can be extracted from the model of the application. Indeed designers can be satisfied by the fact that they are able to verify the complete functionality of the system under test cases; the model of the crane and the control strategies are validated. We expect more for system designers. We consider that such a system-level model must first be able to study system performance properties and then be able to generate most of the real-time prototype.

5.1   Performance studies

In our above simulation we have considered that operations are instantaneous. But execution times of computation algorithms are important for the implementation of real time systems. In addition to profiling the application, one possibility consists in "playing" with the several time attributes of the model: influence of the execution times of operations and of the activation period of the controller on the behavior of the crane. We can also add read and write times on the access of shared-variables. This enables us to simulate sensor and actuator accesses.

More essential, we can study the allocation of several functions on software processors. This is very useful to help Hw/Sw partitioning and to decide on the mapping of the functional solution onto a physical architecture. One first technique consists in annotating functions which might encapsulate the functions to be implemented on a microprocessor with three attributes named 'Concurrency, 'RelativeSpeed and 'SchedulingPolicy. Concurrency represents the maximum number of concurrent operations (1 for a microprocessor), RelativeSpeed allows to study the influence of the CPU clock speed on the system behavior, SchedulingPolicy defines the scheduling algorithm used by the CPU. The second technique consists in describing explicitly the physical architecture and the mapping. Each processor is therefore characterized by the above attributes.

5.2 Prototype generation

When the solution is fully validated at the system level, designers have to develop a prototype or the final product. We take profit from the fact that operations are described in C to generate an executable C program using a real-time kernel (VxWorks) for the software part of the prototype. In the case of the crane system, this means that the ControlSystem function is generated as a multi-task C program. In fact, for this example a real-time kernel (RTK) is not needed because there is only one function which is sequential. Therefore we are also able to generate the solution without a RTK; the solution is only one task activated by a 10 ms interrupt signal.

For prototype generation, we have developed a specific generator on the same principle as the C++ generator. All our generators are based on a meta-generator enabling us to implement any translation rules [3].

6. DISCUSSION ON THE SLDL OBJECTIVE

Specifying examples as a reference for SLDL candidates is a very relevant initiative. The crane document is well-written because it is self-explanatory and complete. But we would like to raise some discussions in order to help clarify the objective of SLDL.

What is really the goal of a System-Level Design or System Specification and Design language? The first question concerns the level of the language: is-it a language to express a system specification only, a language for design description and its verification, a language for design description and also able to induce one or several implementations? This question can be raised for any language. VHDL was conceived for chip description and verification; now it is used to produce implementations through synthesis. C is mainly used for software implementation; some designers use it for algorithm specification and we hope to be able to use it for hardware synthesis soon.

Our solution - the functional model - is not appropriate to express specifications, but is efficient to describe design solutions, to verify functional behaviors, to study performance properties and to derive implementations.

The second question concerns the meaning of the word "language". What is a language and what is not a language? Figure 6 shows that one of the objectives is to link two worlds: the engineer world, the computer world.

Engineers use notations which are clear, understandable for them. A notation is based on a set of well-defined concepts. Computers use languages which are understandable and executable for them. A computer language is based on or generates a computational model. A notation can be automatically translated into a computational language usable for each development stage: specification, design, implementation. From this figure, our position is clear: engineers need well-defined notations corresponding to their engineering cultures which can be automatically translated into a computer language.

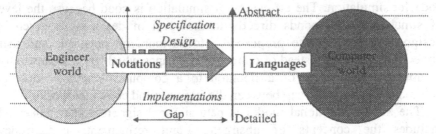

Figure 6. Link between two worlds in system design.

Regarding the MCSE approach, we have defined a graphical notation (the functional model) enhanced with textual annotations (attributes and the algorithms of operations). This mixed notation is automatically translated into an equivalent textual description (MCSE textual model) which is then translated into a standard programming language (C++, C or VHDL). Therefore the functional model can be considered as a candidate for a subset of SLD "language". The graphical model includes functions which represent any kind of active functional components and three types of relationships: synchronization by events, data exchange by shared-variables, message-passing by ports. These three relationships are natural and intuitive for engineers. This model includes the concepts of hierarchy, concurrency, communication and time. A function can be refined by a structure or a behavior. A behavior can potentially be described by any kind of specification models. The result could be a multi-formalism description. We have developed our macroscopic behavioral model specifically for performance modeling and evaluation [2].

The third question concerns the scope of the SLD "language". We consider that a language alone is of few efficiency. This is why we raise this question as follows: Do we search for a language to describe an engineering problem? or do we need a set of notations with underlying concepts useful for thinking on problems and enabling to describe and verify solutions? Or do we need a set of concepts and notations (as above) as well as a seamless related design methodology supported by tools enabling to start from the customer needs and going down to the best appropriate implementation? As a matter of fact, we are in favor of the latter request and our research work on the MCSE methodology has been a long-term attempt to develop a part of the answer which can only be the emergence of several complementary viewpoints and approaches.

7. CONCLUSIONS

In this paper we have demonstrated the usefulness of the MCSE functional notation to describe and validate the Crane system. Our mixed graphical and textual model is automatically translated into a C++ program code for simulation. The efficiency of simulation is good because the level of simulation corresponds directly to the level of descriptions given by designers. We have also proposed to focus more on notations and their underlying concepts instead of computer languages which are a means to use computers. Translation of a notation to a computer language must be automatic to bridge the gap between the engineer and computer worlds.

The MCSE functional model is natural for designers and engineers. It includes the concepts of abstraction and refinement, concurrency, communication, data structuring, time, preemption, scheduling. The tool we have developed automatically translates our notation into a computer language which is dependent on the target implementation (hardware or software). Our approach aims at automatically translating the most difficult part of a system which is the structure of systems (component instantiation and communications between them) instead of focusing on the generation of sequential behaviors which is well mastered today.

8. REFERENCES

1. Calvez JP, *Embedded Real-Time Systems. A Specification and Design Methodology.* John Wiley 1993
2. Calvez JP, 'A System-Level Performance Model and Method', in *Current Issues In Electronic Modeling, Volume 6: Meta-Modeling : Performance and Information modeling,* J.M. Berge, O. Levia, J. Rouillard (eds), Kluwer Academic Publishers, 1996, pp 57-102
3. Calvez JP, Heller D, Muller F, Pasquier O. 'A programmable multi-language generator for CoDesign', DATE'98, Paris, France, 23-26 Feb 1998
4. Foucault B, Calvez JP, Lobao X, Olcoz S, Villar E. 'CoMES: CoDesign Methodology foe Embedded Systems', EMMSEC'99, Stockholm, Sweeden, June 1999
5. Moser E, Nebel W. 'Case study: system model of crane and embedded control', DATE'99, Munich, Germany, 9-12 March 1999
6. Nebel W, Gorla G. ' Java, VHDL-AMS, ADA or C for System-Level Specifications?', DATE'99, Munich, Germany, 9-12 March 1999
7. Pasquier O, Calvez JP. 'An object-based executable model for simulation of real-time Hw/Sw systems', DATE'99, Munich, Germany, 9-12 March 1999
8. Schulz SE, 'The new System-Level design Language', Integrated System Design, July 1998
9. Tutorial on MCSE: http://www.ireste.fr/mcse/htmlan/annoncean.html
   CoMES Web site: http://mcse.ireste.fr
10. SLDL web site: http://www.inmet.com/SLDL

SYNTHESIS

Automatic VHDL Restructuring for RTL Synthesis Optimization and Testability Improvement

Dario Corvino[#], Italo Epicoco[#], Fabrizio Ferrandi[#], Franco Fummi[¥], Donatella Sciuto[#]

[#] Politecnico di Milano, Dipartimento di Elettronica e Informazione, Milano, Italy
[¥] Università di Verona, DST Informatica, Verona, Italy

Key words: Hardware description languages, RTL optimization, Testability.

Abstract: A methodology for modifying VHDL descriptions is the core of this paper. Modifications are performed on general RTL descriptions composed of a mix of control and computation, that is, the typical type of description used for designing at the RT level. Such VHDL descriptions are automatically partitioned into a reference model composed of a controller driving a data-path. We call this transformation "VHDL restructuring". A set of restructuring steps is presented aiming at partitioning any VHDL description while guaranteeing the semantic equivalence of the restructured description with the original one. The main motivation to restructuring is the identification and separation of the two parts (FSM+data-path) which can thus be analyzed by using "ad-hoc" synthesis, testability and design for testability algorithms. Promising preliminary results show that restructuring can sensibly impact on synthesis and testability depending on the size of the isolated control part.

1.   INTRODUCTION

The majority of designers still write their specifications at the RT level even if behavioral synthesis algorithms are well known and commercial behavioral synthesizers are present on the market. The main reason concerns the still low predictability level achieved by behavioral synthesis tools, that makes difficult to estimate at the behavioral level the performance of the automatically synthesized circuit. Behavioral descriptions are translated to a

227

J. Mermet (ed.), Electronic Chips & Systems Design Languages, 227–238.
© 2001 *Kluwer Academic Publishers.*

target RTL architectural model that is usually the finite state machine with data-path (FSMD) [1]. If this kind of description is automatically generated, the controller is clearly separated by the data-path and the data-path is composed of library models (registers, operators, multiplexers). On the contrary, a VHDL description directly written at the RT level is usually a mix of control and computation. In fact, the device behavior is modeled by means of a set of interacting processes and concurrent assignments describing, in a mixed mode, the evolution between states and the computation of results [2].

The main objective of the proposed methodology concerns the translation of a generic RTL VHDL description, representing a FSMD, into a *reference* model. This model clearly identifies the VHDL code representing the controller (FSM) and the VHDL code describing the data-path (D). We call the transformation into the reference model: VHDL *restructuring*. The proposed technique produces a restructured VHDL specification in textual format that can be easily analyzed by a designer to recognize the transformations performed, and clearly identify the controller description and the data-path representation.

The proposed methodology isolates, from the original VHDL code, a controller, which can be easily represented by means of the FSM model (explicit [3] or implicit [4]). In this way, sequential optimization algorithms based on the FSM model can be used such as, for instance: state minimization [5] or state assignment algorithms [6]. Such algorithms are implemented in VHDL RTL synthesis tools, but they are hard to be used since they can be applied only to a design entity that can be modeled by a FSM.

Moreover, the separation of the FSM from the data-path can be used to efficiently solve testability problems of the global circuit. In fact, the testability of the controller can be improved by applying efficient redundancy removal algorithms (e.g. [11]) which assume to deal with circuits modeled by FSMs. In this way, the complete testability of the isolated controller can be reached. On the other hand, the data-path is usually easily testable, if isolated from the controller, due to the removal of many data cycles [12] and due to the isolation of the sequential complexity in the controller. In this paper, we exploit the information derived by the separation of the controller from the data-path to identify scan registers. The scan-chain will include only those registers which allow the controllability/observability of signals interfacing the controller to the data-path. Results obtained by applying this simple strategy are interesting.

The proposed restructuring methodology analyzes VHDL RTL descriptions representing finite state machines with data-paths under the following assumptions:

- The FSMD is described by a single architecture composed of multiple sequential statements (i.e. processes) and concurrent statements (i.e. conditional assignments).

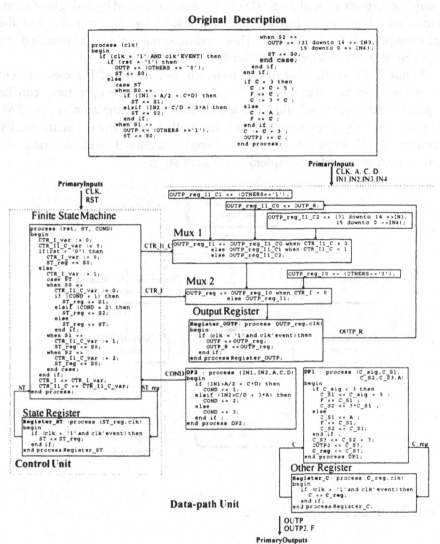

Figure 1. Example of complete restructuring.

- Synchronous descriptions can be based on a single clock signal only. The clock signal is recognized following the typical templates accepted by Synopsys.
- Synchronous or asynchronous reset is allowed.

230

Concerning other VHDL statements (e.g., case, others, functions, etc.), the proposed methodology is able to restructure all RTL VHDL descriptions accepted by Synopsys synthesis tools. Figure 1 reports an example of a complete restructuring of a FSMD (on the top) described by a single process, into a set of processes (on the bottom) describing the controller and the components of the data-path. This restructuring has been obtained by applying all modification steps described in the following.

Figure 2 shows all steps of the proposed methodology. Three concurrent paths are reported after the first step, since such three operations can be applied in any order and allow the identification and extraction of the FSM. The last two steps interconnect the data-path components. Let us briefly summarize all steps of the methodology. Section 3 will describe the corresponding algorithms implementing such steps.

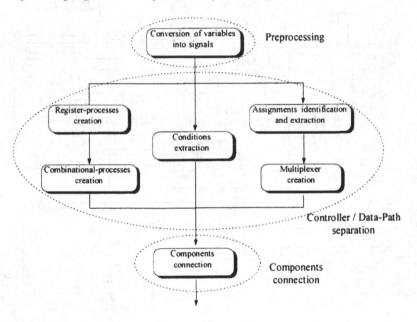

Figure 2. Restructuring flow.

2. VHDL RESTRUCTURING

To simplify the explanation we consider as a running example a FSMD described by a single clocked process. However, the proposed methodology is able to restructure more generic FSMDs described by means of multiple interacting processes. Applying the following restructuring steps to all processes describing the analyzed FSMD can conceptually perform this operation.

Conversion of variables into signals

The main problem concerning the transformation of variables into signals is related to the different semantic of signals and variables with respect to values assignments. In fact, a variable holds the assigned value immediately after the assignment while a signal, assigned into a process, waits for the end of the process or for the first wait to capture the value. For this reason, multiple assignments to the same variable can be modeled only by using multiple signals, each one representing one assignment.

Figure 1 shows on the top-right multiple assignments to the same variable C. The description on the bottom reports a semantically equivalent description, based only on signals, where variable C has been replaced by five signals. This transformation is based on the following algorithm repeated for each variable. Let VAR be the analyzed variable.

1. The VHDL code of the process is partitioned into parts $(P_1 \ldots P_n)$, one for each branch of the code and one for the possibly remaining groups of instructions separated by branches. For instance, the assignments to variable C can be divided into four parts: the code before the if statement (P_1), the true branch of the if (P_2), the else branch (P_3) and the code after end if (P_4).
2. Build a dependency graph for the parts: part P_i depends on part P_j if the execution of P_i is performed at the end of the execution of P_j. For instance, in this example P_4 depends on P_2 and P_3.
3. Set $Cont=0$ and create the new signal VAR_Scont. Set current signal $Curr$ to VAR_Scont.
4. Examine each part P_i following the order of their indices.
   - Set $Cont(P_i) = max(Cont(P_j) \ldots Cont(P_k))$, where $Cont(P_i)$ is the value of $Cont$ for the part P_i, and $P_j \ldots P_k$ represent the set of parts on which P_i depends.
   - Create a new signal for each new assignment of variable VAR and add 1 to $Cont$ at each new assignment. Set current signal $Curr$ to this new signal.
   - If P_i depends on more than one part (e.g., P_j, P_k) and such parts have different $Cont$ values, then add to the parts with lower $Cont$ an assignment to VAR_Smax. In this way the signal used at the exit of a multiple branches selection is referred always by using the same name disregarding which branch has been executed.
5. Finally, every time variable VAR is involved in the right part of an assignment, its name is substituted by current signal $Curr$.

Register-processes creation

Variables or signals requiring registers are identified in the original VHDL description by following a common rule summarizing all synthesis templates that allow synthesis tools to infer registers. A register is created for such signals or variables that are not assigned under all conditions of a selection. For instance, in a clocked process variables/signals are assigned only in the positive (negative) phase of the clock thus constraining the synthesis tool to infer a register to hold the value during the negative (positive) phase of the clock.

This analysis is performed on the original VHDL description that includes variables and signals. Since the previous step has converted all variables into signals, a register is created only for the last instance of a group of signals representing the same variable (see signal C_reg in the bottom-right side of Figure 1). Otherwise, a larger number of registers would be generated and the semantic of the restructured circuit would be different from the original one. A register process is created, for a signal *SIG*, as follows:

1. A new signal ST_sig is created and it replaces all left-hand sides of all assignments to *SIG*.
2. A process is created that is sensible to the clock signal, to ST_sig and, if used, to the asynchronous reset signal.
3. The process assigns ST_sig to *SIG* under the positive or negative phase of the clock.

After this step, separate processes explicitly identify all registers.

Combinational-processes creation

The isolation of all signals representing registers should allow the separation of the original description into a sequential part (the registers) and into combinational parts (the rest). This is not automatically true after the application of the previous restructuring phase. In fact, assume that a register has been inferred for a signal *SIG* since this signal was not assigned in all branches of a conditional selection (see for instance signal ST in the *original description* reported in Figure 1). All assignments to a signal *SIG* have been replaced by assignments to signal ST_reg that represents the input of the inferred register. Since ST_reg is not assigned in all branches of the conditional selection, the synthesis tools will infer another register.

For these reasons, this restructuring step completes the selection branches with simple assignments between output and input of the same register (e.g., ST_reg <= ST;). After this step, sequential and combinational logic is completely separated.

Conditions identification/extraction

Conditional expressions, involving signals with a data size larger than a threshold, must be moved from the controller to the data-path to allow the representation of the controller by using a FSM. This restructuring step is simple since the conditional expression is translated into a separate combinational process that generates a Boolean value that is evaluated in the controller. However, some optimizations can be implemented starting from this simple method.

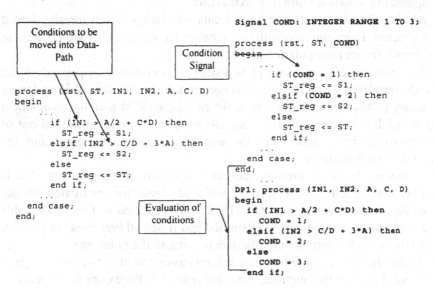

Figure 3. Conditions extraction.

Let us consider the fragment of the process in Figure 1 reported in Figure 3. It includes two complex conditional expressions. A different process could be generated for each conditional expression, but in this case logic synthesis would be performed at first on each single process.

This produces generally worse results with respect to the application of logic synthesis to a more complex process including the common evaluation of both conditional expressions. However, the size of the process must not exceed the maximum optimization ability of the adopted synthesis algorithms. A trade-off can thus be found by working at this abstraction level, before any synthesis step. This operation can be viewed as a *collapse* procedure applied at the gate-level, which merges two modules to optimize their composition.

The inclusion of more than one conditional expression into the same process allows also the reduction of the number of signals, which interconnect the data-path and the FSM. In fact, such signals can be coded and only a logarithmic number of signals becomes necessary. Moreover, the synthesis tool can exploit unused codes to further minimize logic. For instance, a 2-bit COND signal is used to code the three conditions of the example reported in Figure *3*; but three codes only (if, elsif, else) are sufficient. In this way, the synthesis tool has the possibility of using the fourth code as *don't care* conditions, thus resulting in a smaller area with respect to the synthesis of the original process.

Assignments identification and extraction

All signal assignments that have a data size larger than a threshold must be extracted from the controller otherwise its complexity would remain prohibitive for its manipulation as a FSM.

This phase is extremely simplified since the current description does not include variables, given the previous conversion. In this way, it is no longer necessary to take into account the different semantic of a signal assignment and a variable assignment. Thus, signals assignments can be moved out of the controller by guaranteeing the equivalence of the original and the restructured descriptions.

However, the main problem concerns conditional assignments, that is assignments which are performed under a condition evaluated in the controller, for instance, signal OUTPORT_reg reported in Figure 4. In this case each conditional branch is examined and if nested branches are present, the analysis starts from the inner branch moving to the outer one.

1. A new signal is created for each different assignment of the same original signal. Let N be this number of assignments. For the example in Figure 4, N is equal to 3 for the case branch and 2 for the if branch.
2. A new variable CTR is created, with type integer and range (1 to N), for each nested branch. This variable is assigned to a different value (from 1 to N) in each position where the original signal is assigned.
3. Each created variable is connected to a signal (e.g., CTR_I1_C) that selects between the different signals created for the original one. A signal is necessary since assignments are placed outside the original process. Note that, the use of variables and signals is necessary to avoid the generation of unnecessary memory elements in the case the original assignments are executed into a clocked process.

This step of the proposed methodology can be easily evaluated in terms of the complexity of the restructured circuit. In fact, the target of this restructuring method is the minimization of the number of input/output bits of the controller. Thus, the extraction of an assignment is convenient if its data size is smaller than the number of bits necessary to code variables CTR. For instance, in the previous example, the extraction of the OUTPORT_reg signal (of 32bits) removes from the controller 32 input bits and 32 output bits while it adds only 2+1 bits for signal CTR_I1_C and CTR_I, respectively.

Finally, the extracted signals must be correctly synchronized with the selection signals.

Multiplexers creation

The VHDL description of a multiplexer is added to perform the synchronization between the previously extracted signals. Multiplexers are generated with concurrent conditional signals. This is one of the simpler VHDL templates representing a multiplexer that is recognized by all synthesis tools.

Components connection

All previously identified processes are inserted into two design entities, one representing the controller and one representing the data-path. The former is composed of the combinational logic implementing next-state, output functions and registers representing the controller state. The latter includes the rest of the description, that is, combinational processes representing operations between signals, conditional expressions, multiplexers and registers. An architecture interconnecting the two design entities is substituted to the original architecture.

3.   EXPERIMENTAL RESULTS

The proposed restructuring methodology has been implemented in C into the *partition* command of the *COMMIT* environment. The program is based on the LEDA LVS [13] data-base for VHDL parsing, the Berkeley VIS tool for utilities and interface and it is composed of almost 40K code lines. All restructuring steps are directly applied to the LVS data-base and the *reverse* procedure allows generating of the restructured VHDL in a textual format. First, the proposed methodology has been verified for correctness to identify possible semantic mismatches between the original and the restructured VHDL.

We used Binary Decision Diagrams (BDDs) to perform model checking, since the *COMMIT* environment includes the *vhdl2bdd* command that translates (if feasible) any VHDL description into the corresponding BDD based representation. Equivalence checking has been performed on: the original description and the restructured description, the original description and the synthesized restructured description, the synthesized original description and the synthesized restructured description. Complete equivalence has been identified on all examples examined.

Table 1. Synthesis impact of the methodology.

circuit	Original		Restructured					Δ%
	Reg.	Gates	R.Con	R.DP	G.Con	G.DP	Gates	
TX	16	282.4	4	12	190.0	88.0	261.5	-7.4
WRITE	11	154.9	3	8	96.4	48.7	140.3	-9.4
ALPHA	10	336.0	4	6	228.4	41.3	269.0	-19.9
CIDGEN	102	1643.5	11	91	259.4	1322.2	1576.0	-4.1
TOTAL:	139	2416.8	22	117	774.2	1500.2	2246.8	-7.0

The effectiveness of the proposed restructuring technique has been evaluated from the point of view of synthesis and testability. Four industrial VHDL descriptions have been considered; they represent specific TLC circuits mixing control and a data-path operations. Results concerning synthesis are reported in Table *1*. They show the number of registers partitioned between the controller and the data-path, and the area, in terms of equivalent gates, obtained by applying the Synopsys synthesis tool with the LSI10K library. The total number of equivalent gates for the partitioned circuits is lower than the sum of equivalent gates of the control and data parts, since a final stage of optimization of the two parts together has been performed. Ad-hoc synthesis algorithms (states minimization and assignment) have been applied to the controller since its size, after restructuring, has become manageable by such algorithms. Disregarding this optimization, the same effort for synthesis has been spent for the original and restructured versions of the benchmarks. The area of the restructured circuits has been reduced in range 4.1% and 19.9% (7.0% on average). This range is due to the different size of the controllers with respect to the entire circuits. Further improvements can be expected by using different criteria during the separation of the controller from the data-path. In fact, the larger the size of the controller, the higher is the possibility of area optimization obtained through the application of specific algorithms.

Concerning testability, we identified at first scan flip-flops on the gate-level descriptions of the original benchmarks by using an efficient commercial testability analyzer. After partial scan insertion, we generated test patterns by using a commercial TPG. On the other hand, we considered the restructured VHDL descriptions of the benchmarks and we included into the scan chain only those registers allowing the complete observability/controllability of signals interfacing the controller to the data-path. Then, we applied again the TPG to the modified circuits to compute the associated fault coverage. Table *2* summarizes data concerning this comparison. In particular it reports, for each benchmark: the number of faults, the number of redundant faults, the ratio between the number of scan registers and the total number of registers, the fault coverage and the CPU time for test generation.

Table 2. Testability impact of the methodology.

Circuit	Original				
	Faults	Red. Faults	Scan Reg. / Total Reg.	% F.C.	CPU sec.
TX	700	7	7/16	94.4	14.7
WRITE	402	0	6/11	97.3	1.9
ALPHA	1116	24	10/10	97.8	8.3
CIDGEN	4296	20	91/102	96.7	3005.7
TOTAL:	6514	51	114/139	96.7	3030.6
Circuit	Restructured				
	Faults	Red. Faults	Scan Reg. / Total Reg.	% F.C.	CPU sec.
TX	690	0	4/16	94.1	6.6
WRITE	348	0	3/11	96.3	3.1
ALPHA	832	3	10/10	99.6	7.2
CIDGEN	3672	11	12/102	97.1	2652.0
TOTAL:	5542	14	29/139	97.1	2667.2

Results reported in Table *2* show that the fault coverage and the test generation time of the original and the restructured descriptions are comparable for benchmarks TX, WRITE and CIDGEN. However, the number of scan registers of the restructured descriptions is significantly lower than the number of scan registers identified by the testability analyzer on the flattened gate-level descriptions. In the case of the ALPHA benchmark, the same number of scan registers has been selected by the two methods, but the sequential synthesis of the controller has reduced the number of redundant faults thus allowing the test generator to achieve a higher fault coverage.

4. CONCLUDING REMARKS

The paper has presented a methodology for translating general VHDL descriptions representing FSMDs into a reference model where the controller (FSM) is clearly separated from the data-path (D). This transformation, called *restructuring*, preserves the semantic equivalence of the original and the restructured descriptions both before and after RTL and logic synthesis. Synthesis started from the restructured description allows the use of FSM based optimization algorithms which cannot be applied to the original description since it cannot be represented by a unique FSM. Moreover, testability problems can be completely (or nearly completely) solved by connecting into a scan-chain only those registers that allow the controllability/observability of signals interfacing the controller to the data-path. This operation can be performed before any synthesis step, thus avoiding useless recycles in the synthesis flow.

REFERENCES

[1] D.Gajiski, N.D.Dutt, A.C-H.Wu, S.Y-L.Lin, "High Level Synthesis – Introduction to Chip and System Design", *Kluwer Academic Publishers*, 1992.

[2] J.M.Bergé, "Specification of Target Domains and Examples", *Technical Report of Deliverable 1.1.A, Esprit Project 20616.*

[3] G.D.Hachtel, F.Somenzi, "Logic Synthesis and Verification Algorithms", *Kluwer Academic Publishers*, 1996.

[4] O.Coudert, J. C.Madre, "A Unified Framework for the Formal Verification of Sequential Circuits", *Proc. ICCAD,* pp. 126-129, 1990.

[5] G.D.Hachtel, J.K.Rho, F.Somenzi, R.Jacoby, "Exact and Heuristic Algorithms for the Minimization of Incompletely Specified State Machines", Proc. European Conference on Design Automation, pp. 184-191, 1991.

[6] T.Villa, A. Sangiovanni-Vincentelli, "NOVA: State Assignment of Finite State Machines for Optimal Two-Level Logic Implementations", *IEEE Transactions on CAD/ICAS,* pp. 905-924, Vol.9, 1990.

[7] H.Cho, G.D.Hachtel, E.Macii, B.Plessier, F.Somenzi, "Algorithms for Approximate FSM Traversal Based on State Space Decomposition", *IEEE Transactions on CAD/ICAS,* pp. 1465-1478, Vol.15, No.12, 1996.

[8] M.Bombana, G.Buonanno, P.Cavalloro, F.Ferrandi, D.Sciuto, G.Zaza, "ALADIN: A Multilevel Testability Analyzer for VLSI System Design", *IEEE Transactions on VLSI Systems,* Vol.2, No.2, pp.157-171, 1994.

[9] P.Vishakantaiah, J.Abraham, M.Abadir, "Automatic Test Knowledge Extraction from VHDL ATKET," *Proc. DAC,* pp.273-278, 1992.

[10]   A.Ghosh, S.Devadas, A.R.Newton, "Sequential Test Generation and Synthesis for Testability at the Register-Transfer and Logic Levels," *IEEE Transactions on CAD/ICAS,* pp.579-598, Vol. 12, 1993.

[11]   I.Pomeranz, S.M.Reddy, "On Achieving a Complete Fault Coverage for Sequential Machines", *IEEE Transactions on CAD/ICAS,* pp.378-386, Vol.13, No.3, 1994.

[12]   K.T.Cheng, D.D.Agrawal, "A Partial Scan Method for Sequential Circuits with Feedback" ," *IEEE Transactions on Computers,* pp.544-548, Vol.39, No.4, 1990.

[13]   LVS and GRAPHGEN Users' Manual, LEDA, Meylan (France), March 1996.

VHDL DYNAMIC LOOP SYNTHESIS

Marie-France Albenge and Dominique Houzet
ENSEEIHT, 2 rue Camichel, 31071 Toulouse cedex, France.

Key words: VHDL, algorithms, loops, behavioural synthesis.

Abstract: VHDL [1, 2, 3, 4] dynamic loop synthesis is a difficult problem, in the general
             case. It is the reason why no CAD tool implement them.. But in fact, it is
             possible, with specific conditions, to resolve this problem in the particular case
             of logic synthesis because in this context, the size of loops is limited and this
             loops can be explored exhaustively to calculate their maximal size. This first
             step provides a solution to combinational synthesis of loops. Sequential
             synthesis is a second step in order to explore many multi-cycles solutions. A
             loop of N iterations can be treated in 1 to N clock cycles by reducing the
             combinational loop and introducing a counter of cycles. Automatic pipeline
             synthesis is the last step aimed to improve the throughput of the synthesised
             component. Several tools have introduced such possibilities in order to explore
             architectural solutions and compromises between full combinational, full
             sequential, pipelined or not, solutions [5], [6], [7]. The aim of this paper is to
             present the two first steps of loop synthesis in the two first parts. The last part
             will illustrate this approach with a simple algorithm

1.   COMBINATIONAL   DYNAMIC   LOOP
SYNTHESIS

We have considered here that the data types used are from integer type.
This is not a restriction because any type can be transformed, through
coding, in a range of integer.

The two main loop constructs are for loops and while loops. Both can be
synthesised by means of loop unrolling [8].

J. Mermet (ed.), Electronic Chips & Systems Design Languages, 239–247.
© 2001 *Kluwer Academic Publishers.*

For loops follow this structure :
for i in A to B loop....end loop ;
It can be transformed in :
for i in min(A) to max(B) loop
    if i>=A and i<=B then...

with min(A) the left value of A type and max(B) the right value of B type. A and B are the result of dynamic expressions. The size of A and B can be obtained with range calculus if their expressions are only arithmetic and logical ones, that means if there is no condition in their calculus. In that case, the size of A and B can be calculated by propagating the size of the input types in the expressions using A and B. This method points out the important aspect of well defining the data types in order to restrict them to their real range. By this way, it is possible to obtain the range of the result of all arithmetic and logical expressions.

If we consider $R(A)=[A-,A+]$ the range of A and $R(B)=[B-,B+]$, we can find the range of all the results of the VHDL operators using A and B like :

$R(B-A)=[(B-)-A+,(B+)-A-]$
$R(A \bmod B)=[0,\min(A+, (B+)-1)]$
$R(kA)=[kA-,kA+]$ if $k>=0$
$R(kA)=[kA+,kA-]$ otherwise.
and so on for all the operators.

With such operations on ranges, it is possible to propagate the ranges in the direct data dependencies graph of the algorithm to obtain the range of the for loops size. The general case uses dynamic conditions that have to be explored exhaustively, that means we have to consider all the cases of the conditions and to obtain thus all the possible ranges of the final data (A and B in our example).

This step of ranges is a maximal sufficient condition. That means if we take the union of this ranges, we can provide a solution to the maximum size of the loop, that is in fact less or equal than this size. We could improve this result by propagating ranges in the conditions and reducing each range in accordance to the condition. This is possible in the general with set treatment and not range treatment. We have limited at present our study to ranges. Our future work will focus on set definitions and operations in order to improve maximum loop size calculus.

For example, if we consider this condition « if A<4 then », the range of A after the condition is reduced to (A-,4), but with this condition « if A mod 2 = 1 then », the range of A after the condition is (A- +1, A+ -1) when A- mod 2=0 and A+ mod 2=0 else (A-, A+) when A- mod 2=1 and A+ mod 2=1 else (A- +1, A+) when A- mod 2=0 and A+ mod 2=1 else (A-, A+ +1) when A- mod 2=1 and A+ mod 2=0.

With a set and not a range, the resulting set of A would have half the number of values of the initial set of A (even or odd values).

Most of real cases don't require the use of sets to find the maximum size because conditions are often simples and not nested, so the range of the set is sufficient.

The second loop construct is
while condition loop...end loop ;
That can be transformed in :
for i in 0 to N-1 loop
    if condition then...
with N the maximum size of the loop.

The maximal size can be obtained by the same way than for the previous solution, but here, by unrolling the loop and propagating the ranges of the inputs in order to determine the maximal unrolling, that is to say when the condition is always met. If there is no maximal size loop, the unrolling will not converge to meet the condition, and the tool will then detect this non convergence and inform the designer.

The treatment is the following one :

The unrolling of the loop will conduct to define arrays of values (one value per iteration of the loop) for each data used in the loop. Also, the condition can be expressed like a comparison of a data to zero ($c>0$, $c=0$...). Let us consider this example :
R(A)=(0,15)
R(B)=(1,15)
while (A<B) loop A :=A*2...

First step : modify the condition and extract all the dependencies of the condition from the loop, that means eliminate the code that is not linked to the loop exit condition. In our case, we obtain :
while C>0 loop A :=A*2 ;
    C :=B-A ;
end loop ;

Second step : organise each data as an array of values, one per iteration :
A[0] :=A
B[0] :=B
C[0] := B[0]- A[0]
while C(i)>0 loop
  A[i+1] :=A[i]*2 ;
  C[i+1] :=B[0]-A[i+1] ;
end loop ;

Last step : unroll the loop with ranges until the condition is met and verify the convergence.

R(A[0]) :=R(A)

R(B[0]) :=R(B)

R(C[0]) := R(B[0]- A[0])

  := [(B-)-A+,(B+)-A-]

i :=0

if C[0]+ > 0

  i :=1

  R(A[1]) :=R(2*A[0]) :=(2*A[0]-, 2*A[0]+)

R(B[1]) :=R(B[0])

R(C[1]) := R(B[0]- A[1])

  := [(B[0]-)- A[1]+,(B[0]+)- A[1]-]   :=[(B[0]-)-2*A[0]+,(B[0]+)-2*A[0]-]

if C[1]+ > 0

  i :=2

  R(A[2]) :=R(4*A[0]) :=(4*A[0]-, 4*A[0]+)

R(B[2]) :=R(B[0])

R(C[2]) := R(B[1]- 2*A[1])

  := [(B[0]-)- 2*A[1]+,(B[0]+)- 2*A[1]-] :=[(B[0]-)-4*A[0]+,(B[0]+)-4*A[0]-]

Idem if C[2]+ > 0

i :=3

R(C[3]) := R(B[2]- A[2])

  :=[(B[0]-)-8*A[0]+,(B[0]+)- 8*A[0]-]

Considering the previous range definition, we obtain :

R(C[0])=(-15, 14)

R(C[1])=(-30, 13)

R(C[2])=(-60, 11)

R(C[3])=(-120, 7)

R(C[4])=(-240,-1)

The result is N=4 because C[4]<0. Thus, the maximal size of the loop is 4. The propagation of ranges can also be performed in more complex algorithms with conditions as mentioned in the previous part. We are developing an automatic VHDL precompiler that modifies such dynamic loops.

2. SEQUENTIAL LOOP SYNTHESIS

The most difficult part is combinational loop synthesis. The sequential synthesis is only an over set of combinational loop synthesis. This can be done with the splitting of the loops in two parts : one combinational nested loop and one sequential loop. The sequential one is performed with one cycle per iteration. This is a multicycles solution for the implementation of a loop. For example, let us consider the previous case. The synthetizable while loop obtained was :

```
for i in 0 to N-1 loop
if condition then...
```

The sequential one organised with k clock cycles is then :

```
if rising_edge(clk) then
if start_algo then j :=k
elsif j>0 then
for i in 0 to N/k-1 loop
if condition then...
```

The k constant has to be provided by the user as a timing constraint to the CAD tool that calculates it directly according to these constraints.

The pipeline solution, which is similar to software pipelining [8], is another approach to sequential loop synthesis where each iteration or each set of iterations of the loop is designed as a stage of the pipeline, that is to say that the data calculated in the iteration are organised as an array propagated from stage to stage according to the iterations of the loop.

By this way, the loop is also propagating the « start_algo » signal to indicate the validity of the data at the end of the pipeline. The previous example can be expressed as follows, with k2 stages and k cycles per stage (multi-cycles stages) :

```
if (rising_edge(clk)) then
for i in k2-1 downto 0 loop
if start_algo(i) then
j(i) :=k ;
nb(i) :=0 ;
start_algo(i) :=start_algo(i-1) ;
A[i] :=A[i-1] ;
B[i] :=B[i-1] ;
elsif j(i)>0 then
   for u in 0 to N/k-1 loop
   if A[i]< B[i] then
      A[i] := A[i]*2 ;
      nb(i) :=nb(i)+1 ;
```

```
      end if ;
      end loop ;
   else j(i) :=k ;
   nb(i) :=0 ;
   start_algo(i) :=start_algo(i-1) ;
   A[i] :=A[i-1] ;
   B[i] :=B[i-1] ;
   end if ;
```

3.   FULL EXAMPLE

3.1   Algorithm and hypothesis

We take here, as an example to illustrate this method, the typical algorithm used to trace segments of pixels in an image. This algorithm is the following one, written in the C language :

```
Void segment (int xd, int yd, int xf, int yf)
{int dx, dy, s, xinc, yinc, x, y ;

if ((dx=xf-xd)<0)
   { dx=-dx ; xinc=-1 ;} else xinc=1 ;
if ((dy=yf-yd)<0)
   { dy=-dy ; yinc=-1 ;} else yinc=1 ;
x=xd ;
y=yd ;
if (dx>=dy)
{s=-(dx>>1) ;
while (x != xf)
{point (x,y) ;
x+=xinc ;
if ((s+=dy)>=0) {y+=yinc ;s-=dx ;}
}
}
else
{s=-(dy>>1) ;
while (y != yf)
{point (x,y) ;
y+=yinc ;
if ((s+=dx)>=0) {x+=xinc ;s-=dy ;}
```

```
}
}
point(xf,yf) ;
}
```

where the « point » function allows to print a point on a screen.

Before we treat this example, we make the following hypothesis :
-the ranges of x, xd, xf, are from 0 to 639 and those of y, yd, yf from 0 to 479. (like those of a PC screen)
-dx > 0, dy > 0, that is to say the direction of the segment is always from the smallest coordinnates to the highest ones.
-dx > dy, that is to say the segment will be under the diagonal of the image. This simplifies the problem but does not change anything of the result because the two loops of the algorithm are completely symmetric.
With these hypothesis, the loop can be written this way :

```
while (x < xf)
{point (x,y) ;
x+=xinc ;
if ((s+=dy)>=0) {y+=yinc ; s-=dx ;}
}
```

3.2   Combinational loop

If we consider the case of a combinational dynamic loop synthesis, we obtain then the following loop, in VHDL, by transforming the previous one as explained before (we suppose that all the points are put in an array of values) :

```
while (c > 0) loop
  tabx[h] :=x ; taby[h] :=y ;
  h :=h+1 ;
x :=xinc+x;
c :=xf-x ;
s :=s+dy ;
if (s>=0)  then
y :=yinc+y ;
s :=s-dx ;
end if ;
end loop ;
```

When we begin to unroll the loop, the range for x and xf is [0,639] and the range of c is [-639, 639]. At the first step of the unrolling, these ranges become [1, 640] for x and [-640, 638] for c Keeping on unrolling the loop until the convergence is verified, that is until c is negative, we obtain the maximal size of the loop when the range for x is [5639,1278] and the range for c is [-1278, 0], and they are obtained at the 638th step. Of course, because of the symmetry of the algorithm, if we have chosen dy>dx, we would have obtained a maximal loop size of 478.

3.3   Sequential Loop

As we have already said, this loop will be divided in two loops, one nested combinational and one sequential. So, knowing that the maximal size of the loop is 638, and if we want to do the algorithm in k clock cycles, the loop becomes the following one :

```
if rising_edge(clk) then
if start_algo then j :=k; h :=0 ;
elsif j>0 then
for i in 0 to 638/k-1 loop
if c>0 then
tabx[h] :=x ; taby[h] :=y;
h :=h+1 ;
x :=xinc+x ;
c :=xf-x ;
s :=s+dy ;
if (s>=0)  then
 y :=yinc +y;
s :=s-dx ;
    end if ;
end loop ;
j :=j-1 ; h :=0 ;
```

So, to obtain the segment in k cycles, we treat k times N/k iterations simultaneously.

4.   CONCLUSION

The loop synthesis method presented here can be used in a tool dedicated to help the implementation of regular algorithms composed of loops such as a digital image and signal processing. The first results obtained are encouraging. We are preparing a more general version of our method in order to obtain a complete tool making possible the automatic evaluation of many implementation solutions from the full combinational one to the full sequential one. This will be completed with the introduction of sequential operators, that is to say with multicycles or pipelined versions of the VHDL operators (/, *, mod...) inside the loops to obtain full sequential versions. We are studying heuristics in order to limit the field of investigations by the tool according to timing constraints such as bandwidth and latency. The synthesis of one iteration provides a unit time which is used to guide the dimensions of the pipeline. The use of sets of data is also under study in order to provide a more general tool for any kind of loops. This method is also partially teached at ENSEEIHT [9].

5. REFERENCES

[1] R. Airiau, J.M. Bergé, V. Olive, L. Rouillard, « VHDL : langage, modélisation, synthèse », Presses Polytechniques et Universitaires Romandes, 1998.

[2] K.C. Chang, « Digital Design and Modelling with VHDL and Synthesis », IEEE Computer Society Press, 1997.

[3] IEEE Standard VHDL Language Reference Manual, STD 1076-1993, IEEE, 1993.

[4] E.D. Douglas, T.J. Wilderotter, « VHDL Synthesis », Kluwer Academic Publishers1994.

[5] R. Antri-Bouzar, « Du câblage à la micro-programmation : Le micro-programme câblé », Ph.D. Thesis, INPT, France, 1998.

[6] H. Lidonne, J. Elliot (Mentor Graphics), « Explorer tous les choix d'architectures ASIC avant la synthèse », Revue Electronique, N°86, Nov. 1998.

[7] J.N. Contensou, « Le câblage des algorithmes », Ed. Hermès, France 1995.

[8] K. Ebcioglu, « A compilation technique for software pipelining of loops with conditional jumps », 20th Annual Workshop on Microprogramming, Dec. 1987.

[9] D. Houzet, « Formation continue en conception VLSI et FPGA à partir du langage VHDL », CNFM Conference, Saint-Mâlo, 1998.

HIERARCHICAL MODULE EXPANSION IN A VHDL BEHAVIOURAL SYNTHESIS SYSTEM

A.C. Williams, A.D. Brown, Z.A. Baidas
Department of Electronics and Computer Science, University of Southampton, Highfield, Southampton, Hampshire SO17 1BJ, United Kingdom

Key words: high-level synthesis, behavioural synthesis, behavioral synthesis, VHDL, module libraries, cell libraries, hierarchical optimisation.

Abstract: This paper describes a technique developed in the MOODS (Multiple Objective Optimisation in Data and control path Synthesis) behavioural VHDL synthesis system whereby functional data-path modules may be dynamically *expanded* in-situ <u>during</u> the optimisation process, replacing the original 'black box' implementation by its constituent sub-components <u>within</u> the top-level control and data path structure. This enables inter-module optimisation to occur at the sub-module level resulting in substantial area reductions, particularly when targeting restricted device architectures with a limited range of primitive cells, such as FPGAs.

1.   INTRODUCTION

When performing high-level synthesis, the range of low-level functional units (cells/modules) available for implementing operators in the design is an important factor contributing to the efficiency of the synthesised implementations obtained. Synthesis systems typically treat these units as either a single block of combinational logic or a serialised (multi-cycle) 'black-box' implementation, requiring each one to be provided in an optimised technology-dependent form by the target technology library (low-level layout and mapping tools).

J. Mermet (ed.), Electronic Chips & Systems Design Languages, 249–260.
© 2001 *Kluwer Academic Publishers.*

This approach relies on the existence of a wide range of library cells, which is often not available, especially when targeting limited tools for programmable devices (such as FPGAs), which tend to provide only a small range of <u>purely combinational</u> cells.

Instead of manually creating an additional optimised library for each target technology, the MOODS synthesis system [1, 2] uses *expanded modules* to implement complex cells not provided by the low-level tools (mostly optimised topologies and multiple clock-cycle implementations).

2.   BEHAVIOURAL SYNTHESIS

Behavioural, or high-level synthesis [3, 4] is the process of transforming an abstract specification of the *behaviour* of a system, into an equivalent *structural* description that satisfies a set of user *constraints* and *goals* on factors such as final delay and total area. This structure is then processed by logic-level optimisation and layout tools to obtain a final *physical* implementation of the system, based on a particular target technology.

Figure 1 shows the gross design flow in the MOODS synthesis system [1, 2]. This takes as its primary input a VHDL behavioural description, which is first optimised at the language level [5, 6], and then complied into an intermediate form (ICODE) similar to a traditional assembly language. The internal structure of MOODS revolves around a representation of the control and data flow through the synthesised system, ie. a control and data path, which is initially created from a direct translation of the ICODE input forming a naive, maximally serial implementation where each original operation occurs in a single control state, and is implemented by a separate data path functional unit chosen from a library of low-level modules. The synthesis process is formulated around the iterative application of local, reversible transformations to this structure, such as sharing operations within a single data path module, or merging two control states into one. These are applied under the control of an optimisation algorithm, which guides the synthesised implementation towards user-specified objectives on area, delay, power consumption, and 'testability'. Two optimisation algorithms are currently implemented: a heuristic algorithm [2], performing two-way trade-offs between area and delay; and a more general *simulated annealing* [7] algorithm, which is able to perform multi-way trade-offs between any number of target objectives.

Once optimisation is complete, the design is output as a technology-specific gate-level netlist, which is finally processed by low-level logic optimisation and placement/routing (or FPGA mapping) tools [8] to obtain a complete physical description of the synthesised system.

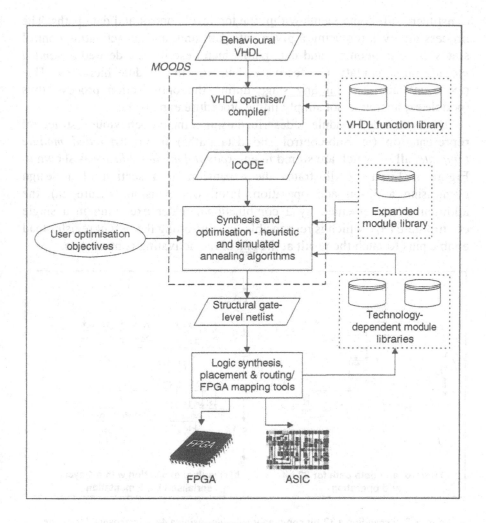

Figure 1. MOODS synthesis system

3. HIERARCHICAL MODULE EXPANSION

Module expansion exploits the inherent hierarchy of arithmetic and logical operators used in module generators such as the FACE environment [9] and BADGE [10]. In these systems, hierarchical module descriptions provide performance models, and guide the generation of module structure <u>independently</u> from the synthesis process. Module expansion, on the other hand, is based on the concept of utilising this hierarchy <u>in the body of the synthesis loop</u> by expanding the sub-structure of a module into its constituent sub-components <u>within</u> the top-level control and data paths. The process involves replacing a given data path unit, and its activating control states, by a sub-control and data path which describe the desired *expanded module* implementation, effectively flattening the module hierarchy. This operation can occur <u>at any stage</u> during the optimisation process, thus module expansion is not simply inline procedure expansion.

Each expanded module is described using a mixed behavioural/structural representation (ie. sub-control and data paths) in an *expanded module template*, all of which are stored in an *expanded module library* as shown in Figure 1. Figure 2 illustrates the expansion of a section of a design comprising a 32-bit add operation. Prior to expansion (Figure 2a), the addition is implemented by a combinational adder executing in a single control state (S_1), which is responsible for controlling the output register load enable pins to latch the result at the <u>end</u> of the activating control state.

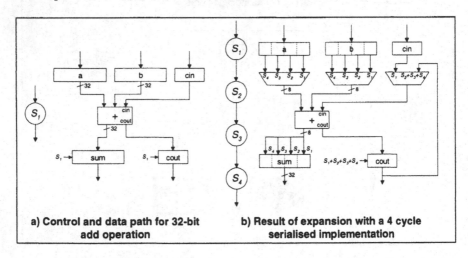

a) Control and data path for 32-bit add operation

b) Result of expansion with a 4 cycle serialised implementation

Figure 2. Expanding a 32-bit combinational adder using a 4 cycle expanded template

Figure 2b shows the result of expanding this adder with a serialised expanded module implementation, based around the accumulation of an 8-bit partial sum over four cycles. Here, the original single control state has been replaced by four new states $(S_1$ through $S_4)$, with the 32-bit adder replaced by a single 8-bit unit, together with a set of multiplexors for selecting 8-bit slices from the 32-bit inputs. Each control state selects a particular set of 8 bits, and writes the result to the corresponding portion of the *sum* register (assumed to have independent load enable controls for each bit), thus building up the complete result over all four cycles.

There are three main factors which influence the effect a module expansion such as this has on a synthesised design:

1. The ability to use sequential, multi-cycle implementations for an operation, when the target library only contains a limited range of combinational cells, facilitates an additional level of trade-off between area and delay without having to create special technology-dependent cells.

2. The hierarchical (and technology-independent) nature of the expanded module description means that additional trade-offs may be made by further expanding sub-modules. For example, the 8-bit adder in Figure 2b might be further expanded into a serial implementation based around a 2-bit adder, thus increasing the total addition time from 4 to 16 cycles. Clearly, the depth of expansion must be balanced against both the total delay, and the cost of the extra circuitry (multiplexors and intermediate registers) required.

3. Because the expanded module sub-graphs are merged seamlessly into the top-level design structure, and are treated identically to all other data and control path elements, further optimisation of the entire design may result in additional improvements through the sharing of sub-modules with other similar data path units. Thus, the area cost of a complex operator may, in some cases, be reduced simply to the overhead incurred by the extra control states, interconnect (multiplexors) and intermediate registers from the expanded module template.

A different approach to the expansion of a 32-bit adder is illustrated in Figure 3. Here, the serialised structure of Figure 2b is unrolled forming a simple, non-optimal implementation, identical to the initial configuration obtained from a behavioural description of the sequential adder.

At first sight this approach would appear to be somewhat more inefficient than before, requiring four separate 8-bit adders and carry registers. It is not difficult to see however, that the simple sharing of all the adders, and all the intermediate carry registers results in a data path structure identical to Figure 2b. This puts the onus on the synthesis system to draw the maximum benefit from the new configuration through further optimisation, which, while not necessarily the quickest and most computationally efficient method, possesses a number of advantages over the earlier model:

1. MOODS will optimise the expanded module in a manner most suited to the target cost objectives. If this requires maximal unit sharing, thus obtaining the original sequential configuration, then so be it. Alternatively, it may be more appropriate to merge the four control states into two, and similarly share the adders, effectively accumulating a 16-bit partial sum over two cycles, as shown in Figure 4. *It can therefore be seen that the most important consideration when designing*

an expanded module is not its structural efficiency, but the range of alternative configurations obtainable through further optimisation.

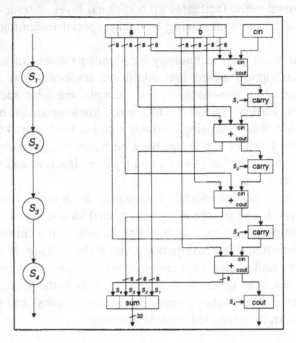

Figure 3. Expanded 23-bit split adder template

2. Feedback loops and unit sharing within the optimised data path of Figure 2b make it much more difficult to optimise the sub-modules in the context of the whole design, since it would first be necessary to unshare the adders and registers. This is particularly significant when using an optimisation algorithm which does not allow design degradations to be performed (such as the heuristic area/delay optimisation algorithm).

3. On the negative side, the immediate effect of module expansion on the overall delay and area will almost always be an increase, thus the expansion process <u>must</u> be viewed as a transformation which increases the <u>potential</u> for optimisation at a later date. Integration of module expansion within an automatic optimisation algorithm is therefore extremely difficult, as it is not possible to evaluate its global effect until synthesis is complete. This problem is complicated by other considerations such as the depth of hierarchy to expand, which units to target, and at what point in the synthesis process should the expansion occur.

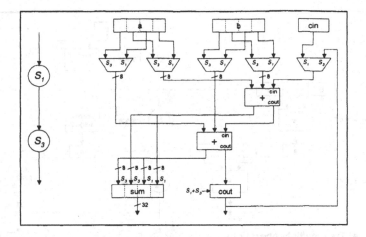

Figure 4. Example optimisation of 32-bit split adder

4.   FURTHER APPLICATIONS

The primary focus when developing the expanded module capability was to provide improvements in area utilisation through the sharing of module sub-components, especially when targeting limited capacity devices such as FPGAs (with a small available set of low-level target combinational cells). In addition to this, module expansion is also used in a number of other ways to provide several additional features, described below.

4.1   Pipelined Modules

Enabling a synthesis system to consider pipelined low-level modules generally requires the development of specialised optimisation algorithms [11] to overlap pipe stage execution according to the required throughput. MOODS contains no special consideration for pipelining, however the use of pipelined expanded modules, in conjunction with the standard optimisation algorithms, produces a similar effect. Figure 5 illustrates a scenario where two multiply operations (Figure 5a) are expanded using a two-stage pipelined multiplier (Figure 5b). Given a suitable cost function assigning equal priorities to both area and delay, optimisation results in the configuration of Figure 5c, which is equivalent to performing the two operations on a single pipelined data path unit, overlapping their execution by one clock cycle.

256

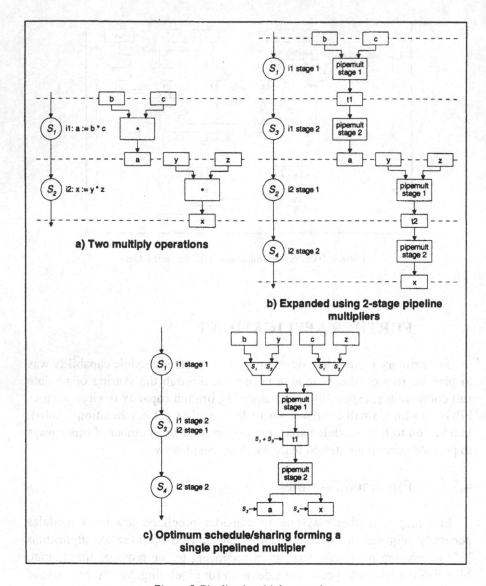

Figure 5. Pipelined multiply operations

4.2 Macro Operators

The applications described so far use an expanded module in lieu of a single low-level library cell, however there is no reason why this cell need actually exist in the target library, so long as module expansion can be guaranteed to occur at some stage during optimisation. *Macro operators* exist purely as expanded modules, accessible from the VHDL source via subprograms defined in a high-level library. During the early stages of synthesis, these are implemented in the data path by a single dummy cell (acting as a placeholder), which is later expanded and the design optimised further. This two-stage approach allows the system to perform a gross level of optimisation while considering the operators as single functional units, before they are expanded and optimised at the sub-module level.

This feature enables the development of technology-independent module libraries, which are automatically optimised by the synthesis process. The fact that these modules are expanded means that highly complex operations (such as floating-point trigonometric functions) need not necessarily occupy vast area of silicon as their sub-modules will be shared throughout an entire design. One such library currently under development supports the IEEE VHDL floating-point standard, providing a large range of functions for real and complex arithmetic.

4.3   Macro Ports

If module expansion occurs after optimisation, just prior to outputting the design as a structural netlist, the expanded control and data path will be directly implemented in the final configuration, thus enabling the local scheduling of operations within the module to be exactly specified on a cycle-by-cycle basis. One of the main drawbacks of behavioural synthesis is the difficulty in interfacing a synthesised design to the outside world, which generally requires the exact specification of operation timing in I/O protocols (eg. memory access control) [12]. By encapsulating the required behaviour in a expanded module with a fixed, pre-defined operation schedule, complex interfaces can be accessed from the VHDL source via a single subprogram call (just like the macro operators described above), thus never requiring the user to worry about any timing details. Macro ports also allow a designer to use alternative protocols (eg. faster/slower memory I/O), by simply changing the expanded module used.

5.   RESULTS

To demonstrate the effectiveness of module expansion for reducing the area of a synthesised implementation, Table 1 lists various figures for three benchmark designs [13, 14]. For each one, the total area of the final hardware implementation is shown (including both control and data paths), together with the total execution time of one complete iteration of the design, and a breakdown of number and size of the main arithmetic units used. Four different sets of figures are presented for each design: *unoptimised*, resulting from the initial configuration prior to optimisation; *optimised*, where a single area optimisation pass is performed with no module expansion; *pre-expanded*, where module expansion occurs prior to optimisation; and *post-expanded*, where a two-stage process is employed, first optimising the design normally, then expanding modules prior to a second optimisation step. In each case, the largest functional units, the multipliers, are expanded using a *split multiplier* expanded module which uses half-size multiply operations, accumulating the result over four clock cycles (ie. long multiplication). Note how the hierarchical expansion process results in 32-bit multipliers first being expanded into 16-bit units, which are themselves later expanded into 8-bit units.

Table 1. Area optimised results[1]

Design	Unoptimised		Area optimised		Pre-expanded area optimised		Post-expanded area optmised	
	Area / Delay	Module usage	Area / Delay	Module usage	Area / Delay	Module usage	Area / Delay	Module usage
FFT	1078 / 11.8	7(+,32) 6(-,32) 4(*,32)	320 / 5.5	1(+/-,32) 1(*,32)	155 / 18.8	1(+/-,32) 1(*,8)	150 / 18.8	1(+/-,32) 1(*,8)
Diffeq	1410 / 5.7	2(+,32) 2(-,32) 6(*,32)	286 / 3.1	1(+/-,32) 1(*,32)	172 / 20.8	1(+/-,32) 1(*,8)	150 / 20.8	1(+/-,32) 1(*,8)
Ellip	571 / 6.6	26(+,16) 8(*,16)	92 / 8.6	1(+,16) 1(*,16)	83 / 6.7	1(+,16) 1(*,8)	68 / 6.7	1(+,16) 1(*,8)

The synthesised implementations are targeted at Cypess pASIC FPGA devices, using the Cypress Warp tools [8] to perform low-level optimisation and device mapping.

The results show a best case area reduction ranging from 26% (ELLIP) to 54% (FFT) when comparing the normal and expanded area optimised results. This is particularly significant for the target devices chosen here, as it makes the difference between a large FPGA (384 blocks) and a smaller one (192 blocks) costing half as much.

Another interesting point illustrated by these results is the difference between pre- and post-expansion. The temptation when performing module expansion is to simply expand all modules prior to optimisation, effectively performing standard inline procedure expansion (pre-splitting). The results show, however, that this approach results in a somewhat less area-efficient implementation than if the expansion is performed part-way through the optimisation process (post-expansion).

[1] Total area is given as the number of FPGA building blocks occupied of the Cypress pASIC FPGA architecture, and total delay is in micro-seconds.

Detailed examination of the synthesised hardware shows that this difference can generally be attributed to the increased complexity of the overall design structure as a result of module expansion, which makes it more difficult for synthesis to optimise interconnects and register allocation due to the increased number of combinations possible. The two-stage approach allows MOODS to first optimise at a slightly higher (and thus simpler) level, and then refine this following expansion. This becomes more significant as the size of the design, and complexity of the modules increases, and is borne out by tests on simpler systems in which the differences between the two approaches become insignificant. The two-stage method also results in a shorter run time for the optimisation process, again due to the reduced complexity of the design structures.

In contrast to the area, the delay figure is largely unaffected by the expansion approach taken. This demonstrates how, for these designs, the process of scheduling operations to control states has far fewer degrees of freedom than that of optimising interconnect and register allocation. Thus the increased complexity resulting from pre-expansion has little impact on the ability of the system to obtain an optimum schedule.

6.   CONCLUSION

The module expansion capability presented in this paper has enabled the existing MOODS synthesis system to perform further optimisation of a behavioural design, primarily as a mechanism for decreasing the area occupied by the optimised structural implementation (at a cost of execution speed). It enables the system to utilise complex multi-cycle modules in situations where the target library only provides a restricted set of simple low-level cells. Furthermore, expanding the hierarchical structure of a module within the top-level control and data paths, allows MOODS to perform inter-module optimisations, further reducing the area requirements.

Expanded modules also enable the creation of a technology-independent module library closely integrated into the VHDL front end, providing the user with a range of features for efficiently exploiting high-level function libraries (such as for floating-point arithmetic) through macro operators, and enabling transparent access to complex I/O protocols (macro ports).

7.   REFERENCES

1.   Baker, K.R. - Currie, A.J. - Nichols, K.G., *Multiple Objective Optimisation in a Behavioural Synthesis System*, IEE Proceedings - G, Vol. 140, No.4 August 1993, pp. 253-260.
2.   Williams, Alan C., *A Behavioural VHDL Synthesis System using Data Path Optimisation*, PhD Thesis, University of Southampton, October 1997.
3.   McFarland, M.C. - Parker, A.C. - Camposano, R., *The High-Level Synthesis of Digital Systems*, Proceedings of the IEEE, Vol. 78, February 1990, pp. 301-318.
4.   Gajski, Daniel D. - Ramachandran, Loganath, *Introduction to High-Level Synthesis*, IEEE Design and Test of Computers, Vol. 11, No. 4, Winter 1994, pp. 45-54.
5.   Nijhar, T.P.K. - Brown, A.D., *Source Level Optimisation of VHDL for Behavioural Synthesis*, IEE Proceedings on Computers and Digital Techniques, Vol. 144, No.1, January 1997.
6.   Nijhar, T.P.K. - Brown, A.D., *HDL-Specific Source Level Behavioural Optimisation*, IEE Proceedings on Computers and Digital Techniques, Vol. 144, No.2, March 1997, pp. 138-144.
7.   Baker, K.R. - Brown, A.D. - Currie, A.J., *Optimisation Efficiency in Behavioural Synthesis*, IEE Proceedings on Circuits, Devices and Systems, Vol. 141, No. 5, October 1994, pp. 399-406.
8.   *Warp User's Guide*, Cypress Semiconductor, April 1996.
9.   Smith, William D. - Jasica, Jeffrey R. - Hartman, Michael J. - d'Abreu, Manuel A., *Flexible Module Generation in the FACE Design Environment*, Proceedings of the IEEE International Conference on Computer Aided Design, 7-10th November 1988, pp. 396-399.
10.   Müzner, Andreas, *BADGE - A Synthesis Tool for Customized Arithmetic Building Blocks*, IFIP Transactions A - Computer Science and Technology, Vol. A-22, 1993, pp. 359-371.
11.   Park, Nohbyung - Parker, Alice C., *SEHWA: A Software Package for Synthesis of Pipelines from Behavioral Specifications*, IEEE Transactions on Computer-Aided Design, Vol. 7, No. 3, March 1988, pp. 356-370.
12.   Gutberlet, P. - Rosenstiel, W., *Timing Preserving Interface Transformations for the Synthesis of Behavioural VHDL*, Proceedings EuroDAC '94, Session V-08, 1994.
13.   Vemuri, Ranga - Roy, Jay - Mamtora, Paddy - Kumar, Nand, *Benchmarks for High Level Synthesis*, Laboratory for Digital Design Environments, University of Cincinnati, 1991.
14.   Panda P.R. - Dutt N., *1995 High Level Synthesis Design Repository*, University of California, Irvine, February 1995.

FORMAL VERIFICATION

Port-Stitching: An Interface-Oriented Hardware Specification and VHDL Model Generation

A.-F. Nicolae, E. Cerny
Laboratoire LASSO, Dép. IRO
Université de Montréal
C.P. 6128, Succ. Centre-Ville
Montréal, QC, H3V 3J7 Canada

Key words: action diagrams, behavioral modeling, timed process algebras, VHDL, pipelines.

Abstract

Design verification is a major concern in the development of Systems-on-a-Chip for which abstract models are required. Standard HDL languages are not well suited for this task. We developed a method for specifying behavioral models of systems as seen from the interfaces - Hierarchical Annotated Action Diagrams, and provided it with formal semantics based on a timed process algebra. HAAD models can be compiled into behavioral VHDL processes that for verification by simulation. The original HAAD formalism had a number of modeling limitations. We describe here a modification in the composition rules of HAAD that gives more flexibility for modeling of pipeline behaviors. The composition is done on a port-by-port basis, by "stitching" together the end of one port behavior with the beginning of the next one, without awaiting the end of the entire enclosing action diagram as in the original method. The semantics of HAAD were formally specified and we established a mapping between the VHDL process generated by our software and the semantic rules.

1. Introduction

Design verification is one of the major problems in System on Chip designs using reusable IP blocks. Abstract models of IP blocks dealing with different views of the components are needed to carry out effective verification by formal and simulation means [10][11].

We developed Hierarchical Annotated Action Diagrams (HAAD) [5] to allow the construction of interface-oriented specifications and VHDL bus-functional models. The leaf-level objects of the HAAD hierarchy are inspired by Timing Diagrams, however, they were generalized to include ports of any (VHDL) type and procedures and predicates annotations attached to actions. Cyclical branching behaviors can be created by using hierarchical operators. The operators are inspired by composition operators from process algebras: Sequential concatenation, Parallel composition with causal rendez-vous communication, Delayed choice, Loop (iteration), and Exception handling. To simplify the composition operations, all Action Diagrams had to be encapsulated between a single *Begin* and a single *End* virtual action. This limits the modeling power, especially in the leaf Action Diagrams (AD, similar to Timing Diagrams), as it requires all activity de-

J. Mermet (ed.), Electronic Chips & Systems Design Languages, 263–272.

scribed by the AD to start after the *Begin* and to end before the *End* action. It is quite usual, due to pipelining of internal operations, that interface "cycles" such as Read or Write are not aligned in time on all ports. To describe such a behavior, it requires decomposing the original cycle description into smaller timing diagrams which then have to be composed using higher-level operators. When the original timing constraints cross the leaf-level boundaries, some ingenuity is required to find the appropriate decomposition into smaller AD's.

In this paper we describe a generalization of the encapsulation rules of Action Diagrams called *port stitching* that allows to specify leaf-level behaviors with activity skewed in time on all the ports. There are no global *Begin* and *End* virtual actions; each port behavior is encapsulated independently by a pair of such actions. In hierarchical compositions the behaviors are connected independently on each port by "stitching" together (merging) the appropriate Begin and End actions on each port[1].

We have specified the Port-Stitched HAAD method using the operational semantics of the $ACTC^P$ [1] process algebra and used its rules to guide and visually verify the implementation of a generator of a behavioral VHDL process that is behaviorally equivalent to the specification of a system given in the Port-Stitched HAAD language. Timing Diagrams have been the inspiration of a number of formal specification system for interface behaviors, e.g., [7][9][11], and as input to synthesis tools for interface controllers, e.g., [8]. However, none of these systems provides the same richness of composition operators and timing constraint types and/or can generate a behavioral VHDL model that implements the formal semantics of the language. Our VHDL model verifies that all input signal transitions satisfy the *assume* timing constraints and produces outputs according to the specified *commit* timing constraints.

The text is organized as follows: in Section 2 we give a simple example of a pipeline and show how its behavior can be described using our method. We also give intuitive semantics of the composition operators. In Section 3, we outline the underlying timed process algebra and its formal semantics, and in Section 4 we describe the generic VHDL process generated from a Port-Stitched HAAD specification. We conclude in Section 5 by indicating possible directions of research.

2. Example and intuitive description

Behavioral model of a pipeline: Suppose the we wish to construct a behavioral model of a pipeline that computes $y = x^2$. The pipeline has the insertion period T

---

1. The name was inspired by the "corner-stitching" technique used in the layout software "Magic" developed by John Ousterhout some time ago.

and carries out the addition in 4 stages, i.e., the latency is 4T. It has one input port X and one output port Y, both of type REAL.

One leaf-level Action Diagram is needed to describe the complete processing of one operand (Figure 1). Its equivalent in the Port-Stitched HAAD lan-

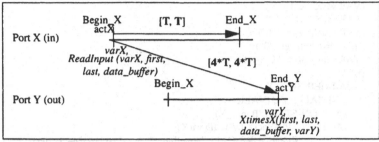

Figure 1: Graphical form of a leaf AD of a pipeline

guage is shown in Figure 2. In Figure 1, the horizontal grey lines represent the time lines of the port behaviors. Each port behavior is delimited by its Begin_P and End_P actions (thin vertical bars across the time lines). The Begin_X action coincides with a real port X input action actX that marks the arrival of a new input value on that port. The End_X action is separated from actX by exactly T units of time (the interval [T, T]). The Begin_Y action on port Y is placed so as to precede actY (which coincides with End_Y) by any amount. The timing of actY is determined by the timing constraints from actX of exactly 4*T units of time. Actions actX and actY are annotated by variables and procedures. Variable varX is automatically assigned the value of port X when actX occurs and the procedure ReadInput is called with the parameters as indicated. The procedure stores the value into the tail end of a FIFO buffer for use by the XtimesX procedure that is attached to actY. When actY occurs, the value of varY is computed by the procedure XtimesX and assigned to port Y. The procedure extracts the value of X from the head of the FIFO buffer. All variable declarations are static. That is, when the Action Diagram is placed into the scope of a recursive composition RecX, new instances of this behavior are dynamically created, but all share the same instance of the variable declarations. To form a pipeline behavior from this description of how one datum passes in time through the operation, imagine that we concatenate an indefinite number of these behaviors into a continuous timing diagram, as shown in Figure 3. Actions are numbered by their occurrence index. The same effect is obtained by using the construct *RecX(Concat (Pipe_F, X))*. The complete hierarchical declaration in the HAAD language is in Figure 3. The syntax is the same as in [5], but in place of a *Loop*, the Port-Stitched HAAD uses recursion over concatenation.

```
(DEFBEHAVIOR Pipe_F
 (PORTS
  -- string enclosed in " " are processed only by the VHDL compiler
  (PORT X IN "inout_bus" MESSAGE)
  (PORT Y OUT "inout_bus" MESSAGE))
 (VAR varX "real" "0")
 (VAR varY "real" "0")
 (VAR last "integer" "0")
 (VAR first "integer" "0")
 (VAR data_buffer "data_table")
 (LEAF
  (CARRIER-SPEC X
   (INITIAL-SPEC (VALID))
   (ACTION-START 'Begin_X)
   (ACTION-SPEC 'actX (VALID varX)
        (PROCEDURE-CALL "ReadInput" varX first last data_buffer))
   (ACTION-END 'End_X))
  (CARRIER-SPEC Y
   (INITIAL-SPEC (VALID))
   (ACTION-START 'Begin_Y)
   (ACTION-SPEC 'actY (VALID varY)
        (PROCEDURE-CALL "XtimesX" first last data_buffer varY))
   (ACTION-END 'End_Y))
  (PRECEDENCE 'actX 'End_X (CMIN 5) (CMAX 5) (INTENT COMMIT))
  (PRECEDENCE 'actX 'actY (CMIN 15) (CMAX 15) (INTENT COMMIT)))  )
```

Figure 2: Stitched-Port HAAD language equivalent of Figure 1.

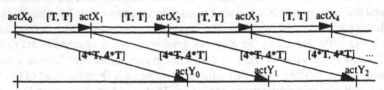

Figure 3: Pipeline behavior from indefinite repetition of that of Figure 1

The first argument of *Concat* can be any Action Diagram and the second argument must be an empty leaf Action Diagram defined over any dummy port. The recursion is unfolded by one instance of the first argument each time the *Begin* action of the concatenation (i.e., *Begin* of the first argument) occurs on any port. We complete now the intuitive description of HAAD.

Leaf AD: Each port behavior is specified as a sequence of actions, enclosed between a pair of virtual *Begin* and *End* actions. The relative positions in time be-

Activated(Term): becomes true when *Term* executes the first action (*begin*) on all its ports.

Wait$_t$*(Term)*: *Term* can wait until t with all its actions.

The syntax of the leaf Action Diagram *Basic_Term* is the same as in [5], except that *Begin* and *End* appear on all ports rather than one pair for the entire AD. The *RecX* operator is restricted to having the free variable X (an empty AD) only as the second term of *RComp* or *Concat*, nesting is allowed.

The details of all the rules of operational semantics can be found in [1][3]. For illustration and for explaining how the VHDL code is constructed, we discuss the rules for RComp (Figure 5) and RecX (Figure 6).

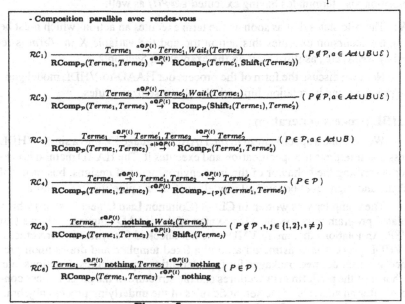

Figure 5: Semantic rules for *RComp*.

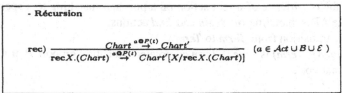

Figure 6: Semantic rule for *RecX*.

RComp:
• If a term can execute an action on a non-communicating port and the other term can wait, the first term executes the action and evolves, while preserving the parallel composition (Rules RC_1 et RC_2).
• If the terms synchronize on *a/b* then both evolve and the composition is preserved (RC_3). If the actions are *end* (RC_4), the comm. port is deactivated.
• If a term completes its execution (RC_5) following $\varepsilon@P(t)$ on a non-communicating port, the composition also executes this action and is replaced by the waiting term shifted until after *t*.
• If the two terms complete their execution by synchronizing on $\varepsilon@P(t)$, the composition also terminates having executed $\varepsilon@P(t)$ as well.

RecX: The rule states that as soon as the term executes an action (which must be *Begin*), the recursion executes this action too and the variable X in *Term* is replaced by *RecX*, etc., as illustrated on the example in Section 2.

Next we discuss the form of the process our HAAD-to-VHDL model generator produces and its relationship with the above semantic rules.

4. VHDL process generation

We compile a specification in the HAAD language [3] into a single VHDL process that interprets the specification and executes it. The HAAD method is useful for describing the behavior of the user environments of designs, bus protocol checkers, and high-level behavioral models such as the pipeline in Section 2.

The compiler was written in CLOS (Common Lisp Object System) which facilitated programming, because the syntax of the HAAD language follows that of LISP. Annotations that are in VHDL are not checked for syntactic correctness. The VHDL process is constructed around a fixed template and draws upon predefined and user defined packages for type definitions, procedures, functions, etc. The choice of the predefined procedures and the structure of the process were constructed after an analysis of the semantic rules of the underlying process algebra.

We first extracted common primitive operations and predicates that appear in the hypotheses or in the consequents[1] of the semantic rules. These include:
• detection of the occurrence of a real action on a port and matching to a specified action in the AD's, handling of *Begin* and *End* actions,
• execution of a transition from *Term* to *Term'*,
• predicate *Empty(P, Term)* is *true* if *Term* completed execution on port *P* or it is not defined on that port,

---

1. The VHDL model has n-ary versions of $ACTC^P$ RComp, DChoice and Concat.

- matching of communicating actions using the set of communicating ports (P),
- predicate $Wait_t(Term)$ detects if $Term$ can wait until t,
- operator $\mathbf{Shift}_t(Term)$ restricts $Term$ to execute after time t,
- term **nothing** indicates the end of execution of a term,
- predicate $Activated(Term)$ is true when $Term$ executes *begin* on all its ports.

To help to structure the implementation and maintenance of the code that executes the rules, we classified the rules into 6 categories, each implemented using the same or similar code, depending on the executed actions, their influence on the dynamic structure of the HAAD hierarchy (e.g., the creation of new instances of AD's) and on the state (active or inactive) of AD's and / or ports:

1. Rules that lead to the activation of AD's and ports at the time of the HAAD hierarchy initialization. E.g., RC_1, RC_2, RC_3 in Figure 5.
2. Rules that lead to the deactivation of ports (and eventually to the activation of ports and AD's in the HAAD hierarchy). E.g., RC_1, RC_2, RC_4, RC_5, RC_6.
3. Deactivation of AD's by executing *end* (the deactivation of ports in a leaf AD may coincide with the deactivation of the enclosing AD). E.g., RC_5, RC_6.
4. Rules that do not modify anything in the hierarchy (no new instances of AD's are created and AD's are not activated or deactivated); they correspond to the execution of actions $a \in Act$ in the leaf AD's and thus the rules are independent of the type of the hierarchical operator. E.g., RC_1, RC_2, RC_3.
5. The recursion rule modifies the hierarchy by creating new instance(s) of the AD either at the initialization of the hierarchy (category 1) or following the deactivation of ports described by category 2.
6. Rule $\mathbf{DC}_2$ (Delayed Choice) handles input actions $a \in In$: the terms that cannot execute the action are deactivated.

The internal representation of a HAAD model inside the VHDL process follows the tree structure of HAAD, but it includes attributes for identifying the state of the system. During simulation, output actions are produced on ports by assigning the specified values to the VHDL signals. The detection of actions and the state updates are triggered when an action is observed on a port. The VHDL process thus awaits transactions or events on all ports. The child nodes in the tree data structure are ordered as the arguments of Concatenation and Exception. An update of the data structure is carried out using a postorder traversal when the process is resumed by either an event or a time-out. Time-outs activate the generation of output actions and detect input actions that arrive outside the latest expected occurrence times as specified by the *assume* constraints.

The VHDL simulation of a leaf AD begins by the virtual *Begin* action and then continues with the specified actions in the order given by the waveforms and the timing constraints. An error in a leaf AD is declared if the initial specified port value is not respected, the assume constraints on input actions are not satisfied, an awaited action does not occur, or if the annotated predicate on an action evaluates *false*. If an error is detected, the execution of the AD is abandoned and the error is propagated up the hierarchy.

The global simulation algorithm proceeds as follows:
- The VHDL process is resumed by a signal transaction or a time-out.
- The detected real actions (if any) are matched with the specified ones on the corresponding HAAD ports or/and time-out is associated with a specified action that should have occurred but did not.

Once all (real, specified) action pairs are identified, they are executed. This consists of recording the occurrence time, updating the expected occurrence time intervals of other actions that are the sinks of timing constraints from the detected action, updating the pointers to the next specified actions on the ports, propagation of errors, etc. If no error is detected and the leaf AD executes *End* on all ports, it completes successfully and is deactivated.

The execution of hierarchical action diagrams follows the rules of $ACTC^P$. These rules only indicate what should be done under certain conditions, i.e., when an activity occurs that does not correspond to any rule, this represents an error and it is processed as the errors described above for the leaf AD's. The overall structure of the VHDL process is as shown in Figure 7. It consists of two parts:
1. Construction of the data structure of HAAD hierarchy and its initialization,
2. A loop implementing the observation of signal transactions and time-outs, updating the signal values and the data structure (the state of HAAD).

The primitives and the semantic rules are implemented in the above procedures and relate to the detection of actions and mapping to specified or virtual actions, the assignment of occurrence times, the evaluation of enabling predicates of rules, the detection of hypotheses and matching with rules, and the update of the state of the hierarchy (for details see [3]). Since actions can occur simultaneously in the leaves of the hierarchy, the evaluation is carried out possibly many times when the VHDL process is resumed.

5. Conclusions

We described an extension to the interface-oriented specification and modeling method based on Hierarchical Annotated Action Diagrams. It provides more flexibility for describing the behavior pipelined systems. This is achieved by composing Action Diagrams on a port-by-port basis using "port stitching". We

```
VARIABLE time_out: time := time'HIGH;
VARIABLE real_actions, out_actions, end_actions, ptr_end_action: pointer_spec_action;
Begin -- PROCESS; "root" is the root of the tree
    load(root); -- creation of HAAD data structure, activate ports
    activate_possible_ports (root, candidats);
    activate_possible_diagrams (candidats);
    init_actions_on_port (candidats, root);
    WHILE NOT root.finish AND NOT root.error LOOP
        WHILE ptr_end_action /= null LOOP
            ptr_end_action := end_actions;
            -- process AD's that completed
            find_finished_hads (ptr_end_action, finished_hads);
            terminate_hads (finished_hads);
            -- activate following AD's
            start_ports (ptr_end_action, ptr_end_action, activated, root, candidats);
            activate_possible_diagrams (candidats);
            initialize_actions_on_port (candidats, root);
            update_end_pointers;
        END LOOP;
            --compute latest occurrence time of expected actions
            compute_time_out (root, time_out);
            -- await real actions or/and timeout, ports is a list of signals
            WAIT ON ports FOR time_out;
                -- detect actions on ports and insert them in a list
                catch_possible_actions (root, real_actions);
                catch_generated_actions (root, out_actions, end_actions); --detect out actions
                process_output (out_actions, end_actions, root); --generate out and end actions
                -- handle in actions and validate ocurrence times and value
                process_reals (real_actions, root);
                validate_port_activities (real_actions); -- update state of HAAD
                -- assign port values to attached action variables
                auto_var_side_effects (real_actions);
                -- execute predicates (Boolean functions) and procedures attached to action
                call_action_pred_and_proc (real_actions, out_actions);
                -- assign values to the corresponding signals
                generate_port_activity (out_actions);
    END LOOP;
END PROCESS;
```

Figure 7: Structure of the VHDL process of a HAAD specification

constructed a VHDL model generator that accepts a HAAD specification and produces an equivalent VHDL process that reflects the semantic rules. This allows simplified validation by targeting specific semantic rules and better error reporting. Changes in the rules can be more easily reflected in the VHDL process.

Future research should proceed in the following directions: Completion of the axiomatisation of the port-stitched HAAD process algebra. Verification of properties of such models by a combination of model checking and constraint solving. Construction of synthesizable bus-protocol checkers and the verification of bus controller RTL implementations against HAAD specifications. Some preliminary work in these directions has been reported in [12].

References

[1] A. M. Andreescu-Hilohi, *Axiomatisation d'un langage de description des chronogrammes (ACTC^P)*, Thèse de maîtrise, Université de Montréal, in progress, 1999.

[2] A. Tarnauceanu, *Logiciel pour la vérification par simulation de la spécification de haut niveau de systèmes matériels*, M.Sc. Thesis, Université de Montréal, 1997.

[3] A.-F. Nicolae, *Modèles et Algorithmes Pour La Simulation Des Systèmes A Temps Réel*, M.Sc. Thèse de maitrise, Université de Montréal, 1999.

[4] S. Gandrabur, *ACTC: Une algèbre de processus temporisée pour la spécification et vérification d'interfaces matérielles*, Ph.D. thesis, Université de Montréal, 2000.

[5] E. Cerny, B. Berkane, P. Girodias, K. Kordoc, *Hierarchical Annotated Action Diagrams: An Interface-Oriented Specification and Verification Method*, Kluwer Academic Publishers, 1998.

[6] 23 F. Jin, E. Cerny, H. Hulgaard, *Maximum Time Separation of Events in Cyclic Systems with Linear and Latest Timing Constraints*, Formal Methods in CAD (FM-CAD'98), Palo Alto, November 1998.

[7] C. Delgado Kloos et W. Damm (Editors), *Practical Formal Methods for Hardware Design*, Project 6128, FORMAT, Volume 1, Springer, 1997.

[8] G. Borriello, *A New Interface Specification Methodology and its Application to Transducer Synthesis*, PhD Thesis, University of California, Berkeley, 1988.

[9] A.J. Daga, P.R. Suaris, *Interface Timing Verification Drives System Design*, ACM/IEEE Design Automation Conf. (DAC'97), Anaheim, CA, June 1997.

[10] T. Bolognese, E. Brinksma, *Introduction to the ISO specification language LOTOS* , Computer Networks and ISDN Systems, 14: pages 25-59, 1987.

[11] J.A. Rowson, A. Sangiovanni-Vincentelli, *Interface-Based Design*, ACM/IEEE Design Automation Conf. (DAC'97), Anaheim, CA, June 1997.

[12] F. Jin, E. Cerny, *Verification of Real-Time Controllers against Timing Diagram Specifications using Constraint Logic Programming*, IFIP EuroMICRO, Vasteras, August 1998.

[13] G. D. Plotkin, *A structural approach to operational semantics*, Report DAIMI FN-19, Computer Science Department, Aarhus University, 1981.

Acknowledgments: The work was supported by grants from NSERC Canada, Center of Excellence Micronet and from Nortel; experiments were done on workstations on loan from the Canadian Microelectronics Corporation.

Formal verification of VHDL using VHDL-like ACL2 models

Dominique Borrione and Philippe Georgelin
Laboratoire TIMA – 46, Avenue Félix Viallet 38031 Grenoble Cedex France

Key words: Formal verification, symbolic simulation, theorem proving

Abstract: When a design reaches the register transfer level, essential architectural decisions have been taken; their validation required extensive simulation of the abstract behavioral specifications. We propose to introduce mechanically supported formal reasoning in the design flow, by producing a model of VHDL behavioral specifications in the logic of the ACL2 theorem prover. Written in Lisp, this model is executable as well as subject to symbolic manipulations. We define the semantics of VHDL data types and behavioral-style statements in the logic. We use macros to generate names, function definitions and theorems automatically, by instantiation of model skeletons, while retaining an algorithmic syntactic flavor. This feature is particularly useful to translate VHDL statements into resembling ACL2 macros, so that the logic formalization remains readable.

1. INTRODUCTION

With the industrial emergence of automatic equivalence and model checking software, formal verification is slowly gaining acceptance as a valuable verification tool, complementing simulation and ensuring greater security. Yet, these tools are inherently limited to verifying the correctness of logic synthesis inputs and outputs (bit and bit-vector level), expressed in the so called "synthesizable subset" of one's favorite design language, or the temporal properties of sequential circuits limited to, roughly, one hundred memory bits. They cannot be applied to show that an algorithm performs an expected arithmetic function on a large data path, unless aggressive abstraction is manually performed, which is an activity that involves quite a lot of expert reasoning to prove that the abstraction is sound. This is where model checking must be complemented with theorem proving.

J. Mermet (ed.), Electronic Chips & Systems Design Languages, 273–284.
© 2001 *Kluwer Academic Publishers.*

When a design description has reached the synthesizable bit-vector level, most of the essential architectural decisions have already been taken; their validation required extensive simulation of the more abstract behavioral specifications. The need for a formal verification of early design decisions is recognized [1], and successful proofs of industrial circuits using general-purpose theorem provers have been reported [2,3]. Yet, the systematic use of reasoning techniques is far from generalized, due to the lack of expertise in design teams. A useful intermediate step appears to be the symbolic simulation of specifications formalized in logic, and supported by a mechanized theorem prover, such as PVS[4] or ACL2[5]. J Moore recommends that formal specifications be executable, to allow successive analysis by simulation on concrete data, then symbolic simulation on indefinite data, and finally application of formal reasoning[5]. These three steps are in increasing degree of abstraction, complexity of data manipulation, and security of results; they also are in decreasing number of verified design cycles or machine instructions per second of host execution. It is thus important to have efficient and easy data simulation of the model, to quickly eliminate obvious errors on test cases, as in the usual simulation-based validation method, before spending more time on symbolic execution and formal proofs.

Our work aims at easing the early introduction of this (mechanically supported) three-step methodology to validate the initial stages of a design. The formal validation achievements mentioned above showed that the method is applicable to systems of significant size. What still needs to be done is the integration of this new validation approach in the HDL-based design flow: hardware designers must be confident that what you prove is what you synthesize. They will continue using conventional simulation or programming languages to execute behavioral specifications on large tests. Symbolic simulation and formal proof will then be welcome, if they are convinced that the formal model faithfully represents the simulated description. To this end:

- the production of the formal model of a design from its HDL specification should be automatic, and based on a verified formal semantic definition of the HDL;
- the formal model should be easy to read, and its correspondence with the initial specification should be obvious.

We apply these ideas to VHDL, using the ACL2 theorem prover. This project follows previous studies that defined formal semantics for VHDL[6,7], and provided some formal verification of VHDL packages[8] and procedural descriptions[9] using NQTHM (the Boyer-Moore theorem prover)[10]. Both NQTHM and its successor ACL2 are based on first-order logic with equality and induction, which is sufficiently powerful to model algorithms implemented in VLSI. They have powerful inference rules and heuristic strategies, and proofs are largely automatic, provided the theory in which a theorem must be established contains the "right" lemmas. Difficult proofs require the introduction of intermediate lemmas, and "hints" to the prover, but we believe most of these lemmas can be automatically generated for well identified application areas, such as the modeling of systems behavior at an appropriate design level. These considerations motivated our choice of this family of provers, rather than "higher-order logic" provers, which require manual guidance to obtain any proof.

With respect to NQTHM, ACL2 [11] offers the following advantages:

- Its input language is Common Lisp. A model written in ACL2 logic is thus executable, and efficient (excellent Lisp compilers are available). One can check that the ACL2 model of a VHDL description provides the same results on the same test cases as a commercial VHDL simulator.
- Its primitive data types include integers and rationals, with a large set of operators and pre-proven theorems.
- The use of Lisp macros allows generating names, function definitions and theorems automatically, by instantiation of model skeletons, while retaining an algorithmic syntactic flavor. This feature is particularly useful to translate VHDL statements into resembling ACL2 macros so that the logic formalization remains readable to the authors of the initial VHDL text. The approach described in this paper was inspired by previous work done at Austin, Texas.

In the rest of this article, section 2 shows the structure of the semantic definition of the standard VHDL language and synthesis packages. Section 3 explains the principle of type definitions and usage in the logic. Section 4 gives the elaboration of a process, and shows the evolution of the "machine state" during symbolic simulation. We end with the status of this research.

2. OVERVIEW OF THE MODELING PROCESS

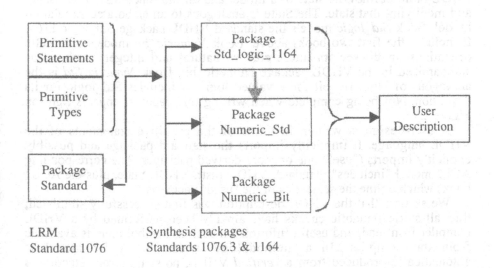

LRM   Synthesis packages
Standard 1076   Standards 1076.3 & 1164

Figure 1. VHDL-1076 primitives, standardized packages and user designs

The development of the formal model in ACL2 closely matches the dependence graph between VHDL standard documents and VHDL descriptions (Fig. 1). The notion of "book" of ACL2 which can be "certified" once and for all, and later be "included" in a model, plays a role analogous to that of "package" in VHDL, which can be "verified" and later "used" to write further packages or circuit descriptions.

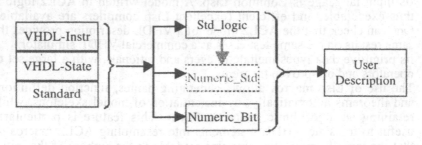

Figure 2: ACL2 books for VHDL language and packages semantics

The modular structure of our semantic definition is shown on Figure 2. Book *Standard* contains the primitive data types and operators of VHDL, which are not already in ACL2. It defines types Bit and Bit-vector, together with all the VHDL operators on these types. Types Boolean, Character, String and Integer are primitive in ACL2: we only add some operators and define the subtypes Natural and Positive. Book *VHDL-Instr* defines the essential declarative VHDL statements (type definitions, signal and variable declarations) and behavioral statements (process, assignments...). Book *VHDL-State* defines the state of a model and all the functions for accessing and modifying that state. The State is analogous to an elaborated simulation model. Book *Std_logic* models the standard VHDL package Std_logic.1164. It includes the first two books above. Book *Numeric_bit* models arithmetic operations on bit-vectors interpreted as natural and integer numbers, as standardized in the VHDL package Numeric_bit. Book *Numeric_Std* is the analogous of Numeric_bit for 9 valued logic. It includes Std_logic for its definition. Not being complete when writing this paper, it appears light on Figure 2.

A user-design is written in terms of the primitive statements of the VHDL language. It implicitly imports the standard package and possibly explicitly imports ("use") one or more derived packages. The corresponding ACL2 model "includes" Standard, VHDL-Instr, VHDL-State, plus the ACL2 books which define the semantics of the "used" packages.

We assume that the VHDL description was first successfully simulated; thus all static semantic checks have already been performed by a VHDL compiler front-end, and useful information such as symbol table is available from the compiler. In a methodology where the ACL2 model is automatically produced from a *verified* VHDL, no static error detection is needed in the ACL2 model, no type or name conflicts exist, and unique naming of all objects can be guaranteed (e.g. for overloaded functions, and variables).

We restrict our discussion to a "behavioral synthesis subset" of VHDL[12], which excludes physical time and non-discrete types. Furthermore, we assume a model to be synchronized by a single master clock, we exclude asynchronous statements (such as asynchronous set and reset), and we consider that all processes have been put in a normal form with a unique wait statement (see [13] for theoretical justification) written: `wait until <clock-edge>`.

3. A READABLE TRANSLATION OF VHDL

The basic idea is to define the logic model for each VHDL statement as a macro, and use the concept of optional key-word in Lisp macros to represent alternatives in the VHDL syntax. We create macros that are syntactically very close to the corresponding VHDL source, so that the user immediately recognizes the translation of his VHDL model, after a few inevitable writing conventions have been learned. The normal pre-processing of the Lisp compiler expands the macros, to produce the formalization in pure Lisp syntax, with the expected nested parenthesized function calls. The result of this expansion is the elaboration of the VHDL source in the ACL2 logic.

3.1 Notations

Due to the idiosyncrasies of VHDL and Lisp, and to avoid conflicts for identical predefined identifiers in the two languages, some notational conventions must be adopted. We chose to add one non alphanumeric character, such as underscore, colon, quotes or star, to keep identifiers and key-words recognizable:

- VHDL key-words that start a new statement are translated as macro names with a leading underscore character. For **instance entity, architecture, type** become **_entity, _architecture, _type**.
- VHDL key-words that continue a statement are translated as optional macro key-words, and take a leading colon. For instance **is, of** become **:is, :of**. Two successive keywords in VHDL are attached by an underscore. For instance, in a type definition, **is array, is range** become **:is_array, :is_range**.
- New identifiers, e.g. the type name in a type declaration, or the object name in a variable declaration, must be written between double quotes. Reference to a previously user-defined type, such as the type name in a variable declaration, is written between stars. For instance, "abc" is the new object being declared, and *bit* is the existing type name in: (**_variable** "abc" **:type** *bit*).

Finally, a minimal amount of Lisp syntax must be obeyed, such as parenthesized macro calls, or the writing of character literals.

3.2 Type definition

VHDL has a few primitive types, and a type definition mechanism to construct new types from existing ones. The primitive types of VHDL that are directly mapped to primitive types of ACL2 are referenced using a normal identifier (rather than enclosed in *), and the constants must obey the lexical conventions of Lisp. All the other types have to be constructed, using a formalized type definition mechanism that declares how the elements of the type are built, and also produces functions and lemmas that will be useful in proving further theorems involving the new type. Table 1 summarizes the type correspondance between VHDL and ACL2.

Comment	VHDL	ACL2
Primitive types in both	Integer Boolean Character	Integer Boolean Character
Primitive sub-types in both	Natural Positive	Naturalp Positivep
Standard type	Bit	*Bit*
Type definition Range Enumeration Constrained array Unconstrained array	Type identifier is Range (e1, e2, ... en) array (dimensions) of array (integer range <>) of	_type "identifier" :is_range :is ("e1" "e2" ... "en") :is_array dimensions :of :is_array integer :range "<>" :of

Table 1: Type identifiers and constructors

The **_type** macro in book VHDL-Instr defines the semantics of the VHDL type declaration statement. It distinguishes enumeration, range and array types, which are identified by a key-word. This macro defines the skeleton for a type definition and what to produce for each category of type. At each call, the macro takes a type name in quotes, e.g. *"name"*, and one or more keywords followed with parameters. It generates a constant for the type:*NAME*, a function that recognizes the elements of the type: *NAME_P*, the "less-than" order function: *ORDER<_NAME*, and one or more theorems which, in particular, guarantee than the recognizer and order functions return a Boolean.

Example 1:

Type definitions in VHDL:

```
type Word_Size is range 0 to 31 ;
type Tab is array (1 to 8, 1 to 16) of Bit;
type index is (ind1, ind2, ind3, ind4);
type tab_index is array (index) of integer ;
type tab2 is array (integer range <>) of integer;
```

ACL2 counterpart of these statements with our macro definitions:

```
(_type "word_size" :is_range (0 :to 31))
(_type "tab" :is_array (1 :to 8 1 :to 16) :of *bit*)
(_type "index" :is ("ind1" "ind2" "ind3" "ind4"))
```

```
(_type "tab_index" :is_array *index* :of integer)
(_type "tab2" :is_array integer :range "<>" :of integer)
```

The key words identify which kind of type definition is being made. **:is** for an enumerated type, **:is_range** for a range (ascending or descending), **:is_array** for an array. Parameters provide the complementary information necessary to define the type, e.g. list (0 **:to** 31) gives the direction and bounds of range "Word_Size".

The macro expansion produces the model of the type definition in the logic. Let us consider type "index". The expansion produces:

- function *INDEX-P(X)*, which returns true if X belongs to type "index". By convention functions ending with P are predicates returning a Boolean value.
- function *order <_ INDEX (X Y)* which returns true if X is smaller than Y in accordance with the VHDL order relation (e.g. ind1 is smaller than ind3).
- the constant list **INDEX* = (8 ("ind1" "ind2" "ind3" "ind4"))* which corresponds to the representation of the type in a symbol table. Type definition kinds are numbered (arbitrarily), and 8 corresponds to an enumerated type, so the functions that manipulate type representations know what should be found in the type definition, here the list of its elements.
- theorems (stating elementary properties to be used in proving further theorems involving objects of the type) which state that:
  1. functions *INDEX-P* and *ORDER<_INDEX* return a Boolean result if their arguments are of the appropriate type
  2. *ORDER<_ INDEX* is a well-formed relation

The function names (*INDEX-P, ORDER<_INDEX*), the type representation name **INDEX**, and the theorem names (omitted here for brevity) are built using the identifier "index" provided as first actual parameter to the macro call, which ensures a systematic naming convention.

The single type call

```
(_type "index" :is ("ind1" "ind2" "ind3" "ind4"))
```

expands into 28 lines of properly indented function, constant and theorem statements which are then accepted by the ACL2 prover. The result is the addition of these definitions to the knowledge base of the prover.

3.3   Object declarations

The book VHDL_State contains all the necessary macros to elaborate a VHDL description in the ACL2 logic. A state is associated to each component of design ; it holds the memory for signals and variables, the program counter of each process and miscellaneous information.

The state is constructed by the macros corresponding to the entity, architecture, port, signal, variable, and process statements. It is composed of two parts:

- A static list of descriptors, analogous to a symbol table, which gives the characteristics of each declared object: (1) identifier; (2) kind, i.e. one among input signal, output signal, local signal, variable; (3) type; (4) initial value.

- A list of dynamic values, initialized according to the declarations, and altered by the execution of the variable and signal assignments. For input signals and variables, only the current value is present in the list. For output and local signals, two spots are reserved in the list: the current and next value of their driver. Since we are only dealing with non timed designs, the full complexity of VHDL transactions can be avoided.

Example 2:

```
entity example2 is
  port (I1 : in bit; I2 : in integer; done : out bit)
end example2;

architecture alpha of example2 is
  signal S1 : boolean;
  signal S2 : index;
  begin
  P1 : process
      variable table : tab_index;
      variable CO : integer := 0;
      ....
```

Assume the definitions of Example 1 have already been processed. The above portion of VHDL text and its ACL2 formalization using the following macro calls are in one to one correspondence:

```
(_entity "example2" :is
  (:port  ("I1" :in *bit*)
          ("I2" :in integer)
          ("done" :out *bit*))
:end "example2")

(_architecture "alpha" :of "example2" :is
  ((_signal "S1" :type boolean)
   (_signal "S2" :type *index*))
  :begin
  ((_process "P1"
    (_variable "table" :type *tab_index*)
    (_variable "CO" :type integer := 0)
    ....
```

These macro calls produce the following list of descriptors:

```
( ("I1"       input     *bit*       #\0)
  ("I2"       input     integer     0)
  ("DONE"     output    *bit*       #\0)
  ("S1"       local     boolean     NIL)
  ("S2"       local     *index*     "IND1")
  ("P1.TABLE" variable  *tab_index* '(0 0 0 0 ))
  ("P1.CO"    variable  integer     0) )
```

where the literals used obey the Lisp denotations: #\0 for character '0', NIL for Boolean False, and '(0 0 0 0) for a list of four zeroes. From these descriptors, the list of dynamic state values is constructed by the initialization function. Each value is retrieved symbolically, using the identifier declared for each object. In the case of an output or local signal declared "name", "NAME" refers to the current value, and "New_Name" refers to the next (driving) value. Thus, the dynamic state for the above example is the list of values for this ordered list of variables:

```
("I1"  "I2"  "DONE"  "S1"  "S2"  "P1.TABLE"  "P1.CO"  "NEW_DONE"
"NEW_S1"  "NEW_S2")
```

The initial value for the state is:

```
(#\0  0  #\0  NIL  "IND1"  '(0 0 0 0 )  0  #\0  NIL  "IND1"  )
```

3.4   Variable and signal assignments

Together with the initial list of values of the (dynamic) state, the correspondence between symbolic name and position in the list, and functions to access and modify one state element are built. Thus, if ST is the current state, with values as above:

`(get_value ST "S2")`	returns "ind1"
`(put_value "ind2" ST "NEW_S2")`	returns a state equal to ST except for "NEW_S2" which has value "ind2".

Functions get_value and put_value are automatically called by the macros which denote variable and signal assignment statements. `(get_value ST "name")` replaces any reference to a variable or a signal declared "name" on the right hand side of an assignment.

When a variable "v" is referenced on the left hand side of an assignment, the result RHS of the right-hand side computation is assigned to the *variable* with `(put_value RHS ST "V")`. When a signal "S" is assigned, the result RHS of the right-hand side computation is assigned to the *driver next value*, with `(put_value RHS ST "NEW_S")`.

Example 3

Assume the declarations and the initial state values of Example 2. Assume the first statements in process P1 are:

```
P1 : process
   variable table : tab_index;
   variable CO : integer := 0;
begin
   S1 <= (I1 =  '0');
   CO := I2 + 2;
....
```

Which are written in our formalization macros:

```
(_process "P1"
   (_variable "table" :type *tab_index*)
   (_variable "CO" :type integer := 0)
```

```
:begin
   ((_<= "S1" '(= "I1" #\0))
   (_<- "CO" '(+ "I2" 2))
   ....
```

_<= is the symbol for signal assignment, _<- is the symbol for variable assignment; #\0 denotes bit '0'. The macro expansions generates, for each assignment, the appropriate call to put_value applied to the current state of the model: each sequential statement (except the first one) thus applies to the state produced by the previous statement (Let* in the generated Lisp code).

Let ST1, ST2... denote successive values of the dynamic state of the model. The two assignments of the example are macro-generated as:

```
(let*
   (ST1 (put_value (EQUAL (get_value ST "I1") #\0)
                    ST "NEW_S1"))
   (ST2 (put_value (+ (get_value ST1 "I2") 2) ST1 "P1.CO"))
   ....
```

Thus, the sequential statements in process P1 produce a succession of dynamic states. Initially:

ST =   (#\0 0 #\0 NIL "IND1" '(0 0 0 0) 0 #\0 NIL "IND1")

After the first statement:

ST1 =   (#\0 0 #\0 NIL "IND1" '(0 0 0 0) 0 #\0 T "IND1")

After the second statement:

ST2 =   (#\0 0 #\0 NIL "IND1" '(0 0 0 0) 2 #\0 T "IND1")

4.   SYMBOLIC SIMULATION

If all signals and variables are assigned with constant values, the model is simulated. Otherwise, the right-hand side of an assignment is kept symbolic and the expressions are transmitted along the successive statements, thus performing a symbolic simulation of the model. The power of ACL2, and the pre-proven lemmas, are used to simplify the symbolic expressions and prove them equal to specified ones.

In an architecture, each process statement is formalized as a function which takes as input parameter a state, and produces a modified state. The function body performs the sequential statement execution, as explained in section 3.4, and the state returned by the last statement is the state returned by the process. Since each process only writes in its local variables and in the new values of its signal drivers, the order in which concurrent processes are called and transmit the dynamic state to each other has no influence on the result.

After all processes have been called, the simulation cycle ends with the computation of the resulting dynamic state, in which all the new signal values are used to compute the current signal value (calling a resolution function if the signal is resolved, or just replacing the current with the new value in the non resolved case).

The simulation function is therefore a recursive function of the following template (expressions are written in infix notation for readability, n is the number of simulation cycles, ST is the dynamic state):

```
(defun simu (n ST)
  (if (n is a natural)
     (if (n = 0) return ST
      else
        (simu (n - 1)
              (let*  ((ST1 (FN_process1 ST))
                      ((ST2 (FN_process2 ST1))
                       ...
                      ((STk (FN_processk STk-1))
                  (updatesignals STk))))))
     nil))
```

where each ((STk (FN_processk STk-1)) calls the function that formalizes process k on the state STk-1 produced by process k-1, and returns a new state STk.

5.   THE PATH FROM VHDL TO PROOF

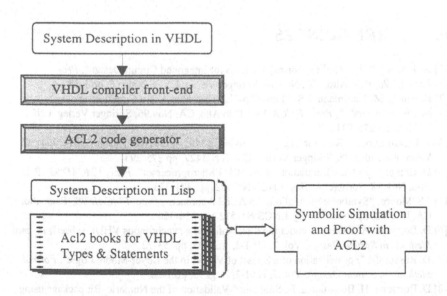

Figure 4. automatic path from VHDL to formal proof with ACL2

Figure 4 depicts our vision of how to obtain an automatic translation of a VHDL behavioral specification into a provable ACL2 model. A system description is first input to an industrial compiler front-end, which performs the syntactic and static semantic checks. We therefore need not worry about those items. From that point the ACL2 macrocode is produced. The macro expansion "includes" the certified books and can be submitted to the ACL2 system for value or symbolic simulation. The power of the theorem prover is

284

then used to demonstrate that the symbolic expressions obtained from the model implement or are equivalent to the expected specified function.

A prototype ACL2 code generator has been written. The difficulty lies in writing macros that not only generate the model in the ACL2 logic, but also generate the useful intermediate lemmas, the simulation function, and the state updating functions, in a way that facilitates proofs of theorems.

Preliminary experiments, performed on synchronous VHDL models, show that the value and symbolic simulations with ACL2 can easily be obtained with our set of macros. This property is an essential first step, to guarantee that the formal model faithfully represents the semantics of the VHDL it derives from. Subsequently, the formal proof that the ACL2 model indeed implements a specified function requires human expertise. Still, we believe that our methodology, which starts from standard design languages, will improve the acceptance of formal methods as an integral part of the design and verification process.

Acknowledgements: The authors are thankful to Vanderlei Moraes Rodrigues for fruitful discussions and helpful comments on a previous version of this paper.

6. REFERENCES

[1] K. Keutzer: "The Need for Formal Methods for Integrated Circuit Design", *Proc. FMCAD'96*, Palo Alto, CA, Nov.96, Springer Verlag LNCS N°1166, pp.1-18

[2] B. Brock, M. Kaufmann, J S. Moore: "ACL2 Theorems About Commercial Microprocessors", ", *Proc. FMCAD'96*, Palo Alto, CA, Nov.96, Springer Verlag LNCS N°1166, pp.275-293

[3] A.J. Camilleri: "A Role for Theorem Proving in Multi-Processor Design", *Proc. CAV'98*, Vancouver, June 98, Springer Verlag LNCS N°1427, pp.275-293.

[4] D. Greve: "Symbolic Simulation of the JEM1 Microprocessor", *Proc. FMCAD'98*, Palo Alto, Nov.98, Springer Verlag LNCS N°1522, pp. 321-333

[5] J S. Moore: "Symbolic Simulation: An ACL2 Approach", *Proc. FMCAD'98*, Palo Alto, CA, Nov.98, Springer Verlag LNCS N°1522, pp. 334-350

[6] D. Borrione, A. Salem: "Denotational semantics of a synchronous VHDL subset", *Formal Methods in System Design*, Vol. 7, N° 1-2, Aug. 95, pp. 53-72

[7] D. Russinoff: "Formalization of a Subset of VHDL in the Boyer-Moore Logic", *Formal Methods in System Design*, Vol. 7, N° 1-2, Aug. 95, pp.7-26.

[8] D. Borrione, H. Bouamama, R. Suescun: " Validation of the Numeric_Bit package using the NQTHM theorem prover", *Proc. APCHDL'96 Conf.*, Bengalore, India, Jan. 1996.

[9] F. Nicoli: "Verification formelle de descriptions VHDL comportementales", Ph.D., Université de Provence, Marseille, France, July 1999 (in French)

[10] R. Boyer, J S. Moore: "*A Computational Logic Handbook*", Academic Press, 1997

[11] M. Kaufmann, J S. Moore: " An industrial strength theorem prover for a logic based on Common Lisp", IEEE Trans. On Software Eng.ineer., Vol. 23 N°4, April 97, pp.203-213.

[12] IEEE Synthesis Interoperability W.G. 1076.6: "Draft standard for VHDL Synthesis Subset Level 2", http://www.eda.org/siwg.

[13] D. Déharbe: "Vérification formelle de propriétés temporelles: étude et application au langage VHDL", PhD, Univ. J. Fourier, Grenoble, 15 Nov 1996 (in French).

SPECIFICATION OF EMBEDDED MONITORS FOR PROPERTY CHECKING

A. Allara[†], M. Bombana[†], S. Comai[‡], B. Josko[*], R. Schlör[*], D. Sciuto[‡]

[†] *Italtel SpA, I-20019 Settimo Milanese (Milano) Italy*

[‡] *Politecnico di Milano P.zza L. da Vinci, 32 – I-20133 Milano, Italy*

[*] *OFFIS – Escherweg, 2 – D-26121 Oldenburg, Germany*

1. INTRODUCTION

The development of new application domains in the telecom market, such as mobile telephony or multimedia services, has led to a substantial increase in the average complexity of single devices. This complexity makes the definition and verification of the interoperability of different modules in a system a very error prone task. Formal verification techniques are becoming more and more attractive as part of the verification methodology, in conjunction to simulation, to guarantee the complete correctness of complex devices. Several examples of applications in industrial environments of these techniques [7,6,1] show that they have reached an acceptable level of maturity. Still, problems of complexity handling imposed by today's model checking technology [8] limit for the moment a wider diffusion.

In some cases the specification of formal properties of digital devices can be very difficult since to prove the correctness of their behaviour it is necessary to verify a complex protocol characterised by long input/output sequences in the model. It is common practice [6] in such cases to implement a *monitor* inside the environment (see Figure 1.a). This monitor asserts a particular variable *monitor_valid* whenever a correct input/output

285

J. Mermet (ed.), Electronic Chips & Systems Design Languages, 285–294.
© 2001 *Kluwer Academic Publishers.*

286

sequence is detected. Having such a variable makes the formal property very easy to express.

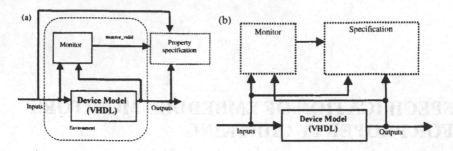

Figure 1. (a) circuit under test with monitor, (b) monitor becomes part of the formal property specification

The main advantages of this technique are the complete independence from the actual implementation of the device under test and a more safe and effective way to specify properties even if the state space increases.

In this paper we will introduce a more general approach based on the idea of implementing monitors directly as a part of the formal specifications (see Figure 1.b). This does not require the setting of the test environment anymore and it allows a clear distinction between the model under test and its formal specification. This is a real advantage, since the embedding of monitors into formal properties could help the *Compositional Verification Approach* where properties of one hierarchical level are proved by composing properties of their submodules through tautology checking. Compositional verification includes all the information of the device into its own set of formal properties. However, if monitors are used, this information is not homogeneous and this could lead to problems both of clearness and integration.

Symbolic Timing Diagrams are used in the following because they are an integral part of the model checking environment used in our site. Anyway the approach does not rely on them and is more general. In fact CTL with temporal variables can be used for the same goal.

The paper is structured as follows. Section 2 presents a very short overview of the Symbolic Timing Diagram; Section 3 addresses the problem of formal specification of HW devices and the use of monitors, while Sections 4 and 5 detail the proposed embedded monitors. Their definition and application in the verification of the VHDL specification of a telecom device are presented in Section 6. Finally, The future lines of work are presented in Section 7.

2. A SHORT INTRODUCTION TO STDS

In the present design approach, we employ the extended Symbolic Timing Diagrams (from now on simply called STDs) [4,3,9,5] as a mean for specifying the behaviour of a digital device. The user defines the properties (expected behaviour) of the device under consideration in a graphical form, which is automatically translated into temporal logic (CTL).

STDs are characterized by the presence of general waveforms, whose events are assertions (any VHDL expression is allowed), and constraints between such events. An example of STD is shown in Figure 2.

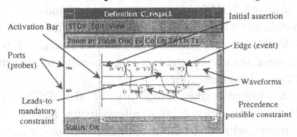

Figure 2. Symbolic Timing Diagram

Two classes of specifications need to be defined: *commitments*, to specify the system behaviour, and *assumptions*, to specify the behaviour expected from the environment for stimulating and responding to the device evolution. When defining the property to be verified, then, a single commitment and one or more assumptions are selected for model checking with respect to the model to be validated. The model checker will prove the commitment by generating all possible input sequences fulfilling the assumptions.

An STD is *activated* when the initial assertions of all its waveforms are satisfied or when the specification imposes that the diagram has to be satisfied only at the beginning of the system run: the activation mode is *initial*, if it is activated only once at the beginning of each system run, *invariant*, if it is activated whenever the initial conditions are met during the system run.

When the diagram is active, the actual behaviour of the associated entity during the system run and the diagram assertions are compared: the actual values of the entity must correspond to the events of the waveform and all the constraints defined between events must be satisfied. This matching process is event-driven: every time a new assertion about the value of the port applies during system execution, if the assertion changes correctly with respect to the waveform and to the constraints specified on that edge, the event is matched, otherwise a violation has occurred.

The severity of the constraint violation depends on the strength of the constraints. A *possible* (weak) constraint indicates that if the target event occurs, it must satisfy the constraint (not necessarily in the specified time limit), otherwise the diagram is deactivated without failure. A constraint with strength *mandatory* (strong constraint) states that the target event must occur, fulfilling the timing specification, otherwise a counter-example is generated.

The expressiveness of STDs is increased by the possibility of using *flexible* and *rigid variables* [2]. *Flexible* variables are logical variables whose value can change in any step. They are modelled as pairs of input ports (of any allowed VHDL type), where the second one represents the value one step delayed with respect to the first one. The two ports are added to the original VHDL device interface and are not connected to any internal module. *Rigid* variables are parameters that initially can assume any value and maintain their values thereafter.

Moreover, to improve the reusability of the STDs, generic STD *definitions* [2] can be adopted, that allow using macros for symbolic waveform assertions, which can be instantiated with particular values when the STD is *declared*.

3.   LIMITATIONS IN FORMAL SPECIFICATION AND USE OF MONITORS

Specification languages used for formal verification of hardware devices allow the description of the behaviour of the circuits modelled, e.g. in VHDL, by means of relationships among the input/output ports. The need of formally specifying and verifying complex sequential circuits from different applications is increasing. Sequential circuits are usually described as finite state machines, where the outputs depend not only on the current input, but also on past values of the inputs. Thus, in general, to formally specify a FSM it is usually necessary to express its behaviour by means of a *sequence* of input/output signals.

The behaviour of a complex device can be triggered by a very long sequence of input/output signals and this leads to write very complex formal specifications. Moreover, in many cases it is not easy to define the state reached after a sequence of inputs using only the input/output relationships.

To clarify the problem, we consider as sample example a regular expression analyser, that asserts the output signal Found when the regular expression ab^+ab is detected on the input port *Data* (Figure 3 shows the FSM recognising this regular expression). Every time an expression is detected, the device starts from the beginning again, i.e. no sequences overlapping is considered.

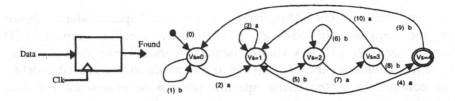

Figure 3. Example of regular expression analyser

In order to verify that the device works properly, we need to prove that:
– for each valid sequence the device asserts the *Found* output signal (from now on called Property 1)
– the *Found* signal is asserted only in correspondence of valid sequences (Property 2).
Property 1 is modelled by the STD commitment of Figure 4.

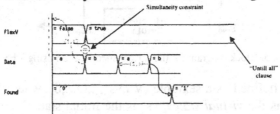

Figure 4. Commitment for Property 1: example of STD with flexible variable

This STD is activated whenever *Data*=a, *Found*='0' and the flexible variable *flexV* is false: after one step the flexible variable is set to true until all the other waveforms/constraints of the diagram have been matched (graphically this waveform ends after all the others). This method avoids matching the activation assertions again when a diagram is active.

However, Property 1 is not sufficient to specify the expected behaviour of the device: it does not say anything about the fact that the Found signal is asserted *only* when this kind of sequence is detected and not in other cases. Currently one way to solve this problem is by using *monitors*. A monitor is a module, generally written in the same language of the design under test (in our case VHDL), that analyses the same inputs (and possibly also the outputs) of the device and asserts a particular variable when a correct sequence of input/output events is detected. Aim of this paper is to provide a method for embedding monitors into the formal specification of properties to be tested.

4. EMBEDDED MONITOR AS VIRTUAL FINITE STATE MACHINE

Our solution to embedding monitors into formal specifications is based on the definition of auxiliary FSM directly specified through STD assumptions. We call this kind of monitors *virtual Finite State Machines* (VFSM) to reinforce the idea that they record all the events needed to define the context, where the formal specification can be proved even if their description is, in every respect, a part of the formal specification.

A VFSMs can read all ports from the interface and provides information about its internal states (called *virtual states*) to the property specification (see Figure 5).

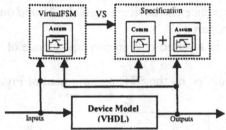

Figure 5. Block diagram for the formal specification using VFSMs

A VFSM is defined as a sextuple $\langle VS,vs_0, I,O,\lambda,\delta \rangle$, where VS is a finite set referred to as the *virtual states*, vs_0 is the initial state, I and O are finite sets referred to as the set of the *inputs* ad *outputs*, respectively, $\delta:VS \times I \rightarrow VS$ is the *next-state function* and $\lambda:VS \rightarrow O$ is the *output function* (Moore FSM). The inputs I are a subset

of (possibly all) the actual input/output signals of the physical device, relevant to the evolution of the VFSM. The *virtual states* VS are, in general, unrelated to the actual states and they may assume a value that can be updated only by the sequence of input signals I. The next state function δ represents how the virtual states evolve with respect to the inputs and present state. The outputs O coincide with the virtual states VS themselves and the output function λ maps one state with the state itself. Hence, actually we need to consider only the quadruple $\langle VS, vs_0, I, \delta \rangle$.

The virtual states of the VFSM can be used in a property specification to be proved in order to recognise particular contexts. In fact, VFSMs, like monitors, can be used to represent in an abstract way the sequential behaviour of a (part of a) circuit, and virtual states can be used in substitution of the internal states in order to identify the right context deriving from a particular sequence of inputs.

5. SEMANTICS AND SPECIFICATION OF VFSM

A VFSM extends the observable behaviour of a system by providing additional information. As there is no transmission of this information to the device model itself (see Figure 5), the VFSM does not restrict the behaviour of the considered model, nor does it modify the behaviour of the device model in any way.

Given a device model M with inputs I and outputs O a property specification *spec* is usually built over the observable interface (I,O). The validity of *spec* for the device M is denoted by $M \models spec(I,O)$.

In an assumption/commitment style *spec* is given as a set of assumptions and a commitment. Verification of such a property is done e.g. by model checking. Virtual finite state machines introduce an additional – expected – observation of the device. Enhancing the specification method by adding virtual finite state machines specifications are given by pairs (*VM, spec*). How is such a property specification verified? As mentioned above, a VFSM VM can be seen as an observer of the interface (I,O) which stores some history information of the observed behaviour in its virtual states. This information can be used in the property specification *spec*(I,O,VS). Hence, verification is performed with respect to the enhanced model M × VM that provides additional information for the property. Formally, we prove the validity of *spec*(I,O,VS) with respect to the composition of M and VM:

$$M \times VM \models spec(I,O,VS) .$$

Hence no modification of the underlying model checking procedure is required. As the transition relation of the virtual finite state machine VM is complete for all possible inputs and no outputs are transmitted to the device M, the behaviour of M × VM restricted to the observable (I,O) is equivalent to the behaviour of M. If the specification spec does not refer to any virtual state the validity of $M \times VM \models spec(I,O)$ is equivalent to the validity of $M \models spec(I,O)$. From this point of view it should be clear that the virtual finite state machine is a kind of memory providing history information for the property specification and does not restrict the behaviour of M.

For example, the behaviour of the device described in section 3 can be defined using the virtual states of a VFSM coincident with the FSM of the regular expression analyser. The two properties to be proved can be efficiently represented by the commitment of Figure 6.

It simultaneously proves both properties, by stating that the output signal *Found* is asserted if and only if the virtual state *VS* is equal to 4. This condition must hold *forever* (in this case the waveform is graphically represented with dashed lines).

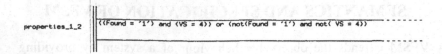

properties_1_2 ‖ ((Found = '1') and (VS = 4)) or (not(Found = '1') and not(VS = 4))

Figure 6. Commitment using virtual state VS to prove Properties 1 and 2

One of the most important benefits of this approach is the possibility of completely embedding a monitor as a list of assumptions into the set of formal properties of a device in a linear and clear way without any need to extend the features of current model checkers.

6. AN INDUSTRIAL APPLICATION

This method has been applied to the verification of a telecom device called ILC16 (16 HDLC Channels Italtel Link Controller), used in the Italtel telephone exchange, a custom ASIC implementation of 16 Channel HDLC Communication Controller with a mixed HW/SW architecture.

Each communication link allows management of data in transmitting (*tx*) and receiving (*rx*) modes at the same time. During the *read* mode, the data are inserted into the data storage memory for a future analysis. The HDLC module receives data from the external environment, runs some OSI layer 2 functionalities (for example the CRC checking) and then transmits them to the FIFO that behaves as a buffer. The DMA transfers the data directly from the FIFO to the main memory through a 32bit BUS. During the *transmit* mode, devices previously described are used in a dual way: the DMA module gets the data from the main memory and inserts them into the FIFO.

To show how the proposed approach has been applied to this device we focus on the Fifo_rx module. This is a synchronous device driven by the two clocks CLK_RX, coming from the environment, and ICLK, the internal clock. The Fifo_Rx contains a buffer of four elements, each one composed of a tag and a data value. The tag value indicates if the data value is an information or a control byte. The DMA device performs read operations, while the HDLC performs write operations. If the Fifo_rx buffer is full, a further write operation generates an "overrun error" (the tag and the data value of the last element of the Fifo_Rx are forced to '1' and to a value representing the overrun event, respectively).

The property we consider here refers to the behaviour of the device in case of "overrun" event occurrence. The automaton of Figure 7 represents the VFSM describing the behaviour of the Fifo_rx in case of read/write operations.

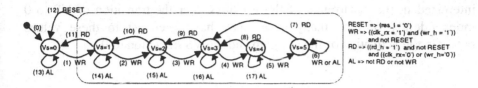

Figure 7. A VFSM of the Fifo rx device

The virtual states change in correspondence with the *read/write* operations. The *write* operation (WR) depends on the assertion of the input signal WR_H and is triggered by the CLK_RX clock. The *read* operation (RD) depends on the assertion of the input signal RD_H, having lower priority than the WR_H signal (to perform a *read* operation WR_H must be disabled) and is triggered by the ICLK that is the reference clock of the device (implicitly considered).

The information of the VFSM is used to prove that when an overrun occurs, the device overwrites the last element of the internal buffer with a code representing this kind of error. A simple way for detecting this fact without adding further port signals, is to look at the values on the dat_out and tag_out ports when an overrun has occurred (VS has become 5) and subsequently the fifo has been emptied (VS=0) without further write operations: in such a way the content of the last element of the internal buffer is available on the output ports. These concepts are efficiently expressed by the commitment of Figure 8.

This STD is activated when the virtual state VS is equal to 5 and for any value on the dat_out and tag_out ports; then, when the virtual state changes to 0, without performing any further writing operation (wr_h port is de-asserted till the end of the diagram), the dat_out and tag_out ports have to report that an overrun error occurred (the *stable condition* at {true} for VS is a compact way to state that all possible values between VS = 5 and VS = 0 are allowed).

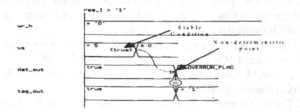

Figure 8. Commitment to prove the overrun property

Note that the output signals DAT_OUT = OVERRUN_FLAG and TAG_OUT = '1' are defined as *non-deterministic points* since we are interested in the occurrence of those values after the transition of VS to 0 independently of how many times they have been set to these values between the activation of the diagram and the VS transition to 0.

7.   CONCLUDING REMARKS

Actual applicability of formal verification strongly depends on the availability of methods and tools allowing the designer to specify the intended behaviour in a compact and effective way. The proposed technique is based on Virtual Finite State Machines that provide the possibility of extracting from the actual description those "*states*", that do not necessarily correspond to the implemented states, necessary for the specification of particular properties. The paper has shown how this embedding of monitors in the formal specification is performed through the instantiations of STDs assumptions. Examples have been provided for a small test case, and for a telecom application. A promising exploitation of this technique is in conjunction with compositional verification and future work will aim to investigate such possibility.

8.   REFERENCES

[1] A. Allara, C. Bolchini, P. Cavalloro, S. Comai, *Guidelines for Property Verification of VHDL Models: an Industrial Perspective*, FDL'98, Lausanne, Switzerland, 6-11 Sept. 1998, pp. 11-20
[2] A. Allara, S. Comai, R. Schlör, *System Verification Using User-friendly Interfaces*, DATE'99, Munich, Germany, March 1999.
[3] W. Damm, B. Josko, R. Schlör, *Specification and Verification of {VHDL}-based System-Level Hardware Designs*, Specification and Validation Methods, ed. E. Börger, Oxford University Press, pp. 331-410, 1995.
[4] R. Schlör and W. Damm. *Specification and verification of system-level hardware designs using timing diagrams*. In Proc. The European Conf. on Design Automation, pp. 518-524, Paris, France, IEEE Computer Society Press, 1993.
[5] R. Schlör, B. Josko, D. Werth, *Using a visual formalism for design verification in industrial environments*, TACAS'98, LNCS, Springer-Verlag, 1998.
[6] J. Y. Jang, S. Qadeer, M. Kaufmann, C. Pixley, *Formal Verification of FIRE: A Case Study*, 34th DAC, proceeding 1997
[7] B. Plessier, C. Pixley, *Formal Verification of a Commercial Serial Bus Interface*, Int'l Conf. on Computers and Communications, Phoenix, USA, pp. 378-382, 1995
[8] M. D. Aegaard, R. B. Jones, C. H. Seger, *Combining Theorem Proving and Trajectory Evaluation in a Industrial Environment*, 35th DAC, San Francisco, USA, pp 538-541, 1998
[9] K. Feyerabend, B. Josko, *A Visual Formalism for Real Time Requirement Specifications*, in Proc. Transformation-Based Reactive Systems Development, ARTS'97, LNCS 1231, pp. 156-168, Springer-Verlag, 1997.

Formal Specification and Verification of Transfer-protocols for system-design in VHDL
Practiced on a Sliding-Window-Protocol

Dipl.-Inform. Olaf Drögehorn, Dr.-Ing. Heinz-Dieter Hümmer,
Prof.Dr.-Ing. Walter Geisselhardt
Gerhard-Mercator-University Duisburg, Germany, Faculty of Electrical Engineering Institute of Dataprocessing, E-Mail: {droege|hdh|gd}@uni-duisburg.de

Key words: Formal verification, formal specification, formal proof, TLA, VHDL

Abstract: In order to meet the requirements of upcoming multimedia applications new transfer-protocols have to be developed. But the design of new protocols is highly complex so the support of design-strategies by formal methods is recommended. On base of today's mostly informal, incomplete, and ambiguous protocol descriptions correctness-proofs cannot be performed. In order to realize a new protocol an implementation in hard- or software has to be designed. The recent way to implement hardware-devices is a description in a hardware description language like VHDL or Verilog. In this contribution we will outline a compositional specification-style using a simple temporal logic of actions, TLA. Based on this description we are able to prove the protocol against a service description by using boolean logic. Corresponding to the TLA-specification we will introduce a VHDL-near dialect of TLA, the translation rules, and the mechanisms which will be needed to transform TLA into VHDL.

1.   INTRODUCTION

In order to be able to transfer information between several computer systems the participating network elements have to agree to a particular procedure or protocol. Such a protocol implements a set of services like setting up, maintaining, releasing connections, transferring, checking and correcting data a.s.o. In modern heterogeneous networks several protocols

295

are in use to transfer the user-specific data. These protocols have to be very flexible in order to meet various requirements according to the user's specifications. Therefore, for today's high-speed applications, new protocols must be developed. Since these protocols are highly complex we recommend supporting the design process by formal methods [11].

The main reasons to use formal methods to describe transfer mechanisms are, on the first hand, the possibility to setup an unequivocal description of a given protocol and, on the other hand, to be able to prove the correctness of the described system. Precondition of a formal proof is a list of services the intended protocol should provide. The proof itself is organized as a number of well defined steps by which the services given in the list are mapped on the used protocol-mechanisms. Due to the designers' carelessness the list of services is missing in most cases. Hence, a proof if the protocol-description correctly represents the required services can't be performed.

Although there are standardized specification techniques like „SDL" [17], "ESTELLE" [15] or "LOTOS" [16] we are using TLA (temporal logic of actions [18]) which is based on logical formulas, because it is not possible to prove the correctness of a protocol description with the above mentioned languages. A specification by TLA is similar to the structure of the OSI-model. A protocol-designer has to specify the service-functions the protocol should provide and all the protocol-mechanisms that are needed to meet the requirements. Since all the mechanisms can be described separately, it is possible to set up a library containing reusable mechanisms [13]. In this way the designer can quickly describe the logical structure of a protocol which is very useful for understanding the interaction between the various mechanisms of that protocol.

The advantage of this approach is that the intended system is uniformly described in TLA both at the abstract level and at the fine-grained level. So based on these descriptions it is possible to prove that the fine-grained system implements the abstract system. The verification may be performed either mechanically based on reachability analysis (e.g., state space explorations [14], model checking [4, 5, 8, 10]) or by symbolic logic reasoning. Using TLA for describing both the service constraints and the protocol mechanisms the proof can be reduced to an ordinary logical implication that the protocol implies the service.

For use in practice a new protocol must be implemented in hard- and/or software. In most cases modern high-speed protocols are implemented in hardware because of the required processing speed. Today's usual way to design hardware-devices is based on a description in VHDL (Very high speed integrated circuit Hardware Description Language [2]) or Verilog. With these languages the developer is able to describe in detail the hardware he wants to produce, with all its needs of timings and signals [3]. The resulting description is a signal-driven model of the designed processes. But the correctness of such a description can only be verified by testing the modelled system in an appropriate simulator. As mentioned before, it is more or less impossible to test or explore all the states of modern high-speed protocols. So the upcoming new strategy of our approach is to translate a verified description of a protocol from TLA into VHDL.

Therefore, we have enhanced the description mechanism of TLA in order to meet some requirements of hardware descriptions. The main idea is based on the advantage that TLA describes all possible trajectories or, in other words, all possible behaviours of the entire system. By translating this system to VHDL the state graph of the desired protocol will be implemented in hardware. Assuming the correctness of the TLA description the designer is able to realize the protocol by a simple translation of state graphs.

2. SPECIFICATION LANGUAGES

2.1 The temporal logic of actions (TLA)

A concurrent algorithm is usually specified with a program. Correctness of the algorithm means that the program satisfies a desired property. We propose a simpler approach in which both the algorithm and the property are specified by formulas in a single logic. Correctness of the algorithm means that the formula specifying the algorithm implies the formula specifying the property.

The logic that we propose for describing and reasoning about systems is the temporal logic of actions abbreviated as TLA [18]. It has been developed for the specification and verification of distributed systems. TLA combines a logic of actions and a standard temporal logic to express a complete system.

In TLA a collection *Val* of values and an infinite set *Var* of variable names are assumed. The logic consists of a set of rules for manipulating formulas. To understand what the formulas and their manipulation mean, we need a semantic. A semantic is given by assigning a semantic meaning *[[F]]* to each object *F* of the logic. The semantic of the logic TLA is defined in terms of states. A state is an assignment of values to variables. States and values are purely semantic concepts and they do not appear explicitly in formulas. So what we express with a logical description is a set of states the system can reach. In terms of standard specification-models we describe state-transition-systems with logical formulas in TLA.

To model states in TLA state predicates, called predicates for short, are used. A predicate is a boolean expression built from variables and constant symbols. The meaning $[[P]]$ of a predicate P is a mapping from states to booleans. We say that a state s satisfies a predicate iff (if, and only if) the predicate can be evaluated to true with the correct value-assignment for the used variables in the state s.

For the description of state-transition-systems we define that a primed(') variable x' denotes the value of x in the next state. An action is a relation between old states and new states, where the unprimed variables refer to the old states, and the primed variables refer to the new states. Formally, the meaning $[[A]]$ of an action A is a relation between states, a function that assigns a boolean to a pair of states s, t. We define the function by considering s to be the old state and t the new state. The pair of states is called an A *step* iff the function for this action evaluates to true.

An algorithm consists not of a single state but of sequences of states. So, if we want to reason about algorithms we have to deal with state sequences or trajectories. To do so by formulas we need a simple temporal logic. A temporal formula is built from elementary formulas using boolean logic and the unary operator $\Box$ (read "always"). For example, if $F1$ and $F2$ are elementary formulas, then $F1 \wedge \Box F2$ is a temporal formula. The semantic of temporal formulas is based on behaviours, where a behaviour is an infinite sequence of states. For the sake of brevity we don't want to formalise the complete semantic here. But based on the unary operator *always* we present the basic concepts of expressing liveness.

With boolean predicates and actions we can express the state space an algorithm could cover. So we are describing all reachable states and assertions that something may never happen because these states aren't reachable. These are the safety properties of a system. The assertion that something does eventually happen is called a liveness property. By temporal formulas we are able to describe liveness properties. For example, if we specify a communication protocol with TLA we need liveness properties asserting that the program keeps going. In order to avoid adding safety properties by expressing the liveness of a system we will separate both descriptions [1]. Therefore we will express liveness in terms of fairness assumptions. Fairness means that if a certain operation is possible, then the program must eventually execute it. For concurrent algorithms we need two types of fairness conditions, weak fairness and strong fairness.

For describing fairness conditions we introduce a new operator $\Diamond$ (read "eventually") which can be derived from the operator *always*. For any temporal formula F let $\Diamond F$ be defined by $\Diamond F = \neg \Box \neg F$. This formula asserts that is not the case that F is always false. With these temporal operators we can now define weak and strong fairness. Weak fairness asserts that an operation must be executed if it is often enough possible to do so for a long enough time. "Long enough" means until the operation is executed, so weak fairness asserts that eventually the operation must either be executed or become impossible to execute. To formalise these definitions, we must define the execution and the impossibility of an execution. So $\Diamond$ *executed* can be expressed by the formula $\Diamond A$ where A is the action describing the operation which has to be executed. For the term *impossible* we can define a predicate *Enabled(A)* which will be evaluated to true iff the operation described by the action A can be executed. So the correct definitions of weak and strong fairness can be written in the following way:

$$WF(A) = (\Box \Diamond (A)) \vee (\Box \Diamond \neg Enabled(A))$$
$$SF(A) = (\Box \Diamond (A)) \vee (\Diamond \Box \neg Enabled(A))$$

Considering all mentioned mechanisms we are now able to describe a system with all safety and liveness properties. A system-formula Φ can be constructed which will completely specify all properties of the desired system.

$$\Phi = Init_\Phi \wedge \Box(N) \wedge WF \wedge SF$$

In this formula *Init* denotes the predicate which expresses the initial state(s) of the desired system. All possible states after the initial one are described by the next state relation N, where N is a conjunction of all actions of the system. With the temporal operator *always* it is denoted, that one of the next-state relations has to be true. Last but not least all fairness assumptions are added to the system formula by adding weak and strong fairness conditions for several actions.

With such a formula we can describe a program at each level of abstraction. Both, the properties or the fine-grained system either can be specified by using TLA formulas. For proving that an algorithm described by the system-formula Φ implements several properties specified by a formula Ψ we just have to verify that Φ implies Ψ with the ordinary boolean logic $\Phi \Rightarrow \Psi$. We don't want to explain the verification of algorithms, because it is not the main topic here. A complete introduction how to prove TLA specifications can be found in [18].

So, with TLA we have explained a simple way of describing algorithms and their properties with logic formulas. Because of the simplicity of the used formulas we are able to prove that a system implements several properties by a single boolean implication. In the following chapters we will enhance the basic TLA for using it to describe hardware-oriented systems and to translate them to VHDL.

2.2   Enhancements for cTLA+

Many system-designers would hesitate to use TLA because of the amount of work that will be needed to become familiar with the logic way of describing. Therefore TLA was enhanced several times to get a description language which is easy to understand even for those people, who are not used to think like mathematicians.

One major improvement was made to add the benefits of compositionality. With the introduction of cTLA+ it was possible to specify several parts of a system completely in separate and compose them afterwards [12]. The main advantage of using compositionality is that all the specified properties and assumptions will hold even after combining several processes to a system.

In cTLA+, specification block describe processes. The syntax of cTLA+ is oriented at programming languages like Modula 2. As an example we refer to the definition of the process *BufS* in Fig.1. This process specifies a buffer, which stores messages that were passed from the outside.

```
PROCESS BufS
BODY
    VARIABLES                                    -- Private variables of the process
        var : [[ub, lb : Int;   array : [Int :mapsto: m :in: message_typ]]];
    INIT ==
        var.ub = 0 ∧ var.lb = 0;                 -- Initial State
    ACTIONS
        NReq(m : message_typ) ==                  -- Action Network Request (NReq)
        Between(var.lb, var.ub, var.lb + used_window)
        ∧ var.array' = [var.array, var.ub :mapsto: m]
        ∧ var.ub' = (var.ub + 1) MOD window;
END
```

Figure 1: Process BufS

The state space of a process is modelled by local variables. A predicate headed by the construct *INIT* defines the set of initial states. In addition to the actions defined in the corresponding section, a cTLA+ process may also perform the so-called stuttering step, the execution which does not change the process state at all. For the stuttering step a pseudo-action with the name *stutter* is always implicitly defined.

With respect to separate safety and liveness properties, the process shown here specifies only safety properties [1]. Thus, *BufS* tolerates state sequences, where after a finite number of state changes only stuttering steps are performed. To rule out such state sequences, process actions can be attributed with fairness assumptions like we have shown it above. In the following examples we will concentrate on the safety properties in order to explain the pure mechanism of translation. It can be shown that the VHDL implementations will fulfil the specified liveness assumptions.

In cTLA+ processes are combined to systems similar to LOTOS [16]. The processes interact via synchronous joint actions that are conjunctions of the participating process actions. The transfer of data between processes is done with the action parameters. The vector of the process variables forms the overall system state and the system transitions are described by the joint actions in a separate system module. Because the semantic of cTLA+ is the same as for TLA, we can reduce a cTLA+ description to a plain TLA specification, by setting up the system formula Φ for the composition of several processes. So cTLA+ is an easier way of specifying systems that are composition of subsystems, with logical formulas. In the following we will always assume the use of cTLA+ and the terms TLA and cTLA+ will be used synonym.

2.3   Very high speed integrated circuit Hardware Description Language (VHDL)

VHDL is a language for describing hardware or hardware-near systems. With this language a developer is able to specify in detail the hardware he wants to produce, with all its needs of timings and signals [3]. The resulting description is a signal-driven model of the designed processes.

VHDL has been standardized by the IEEE as IEEE standard 1076 in 1987, and was revised in 1993 [3]. This is a great advantage, since it allows for an independent standard of documentation, as well as for reusing technology. With the development of new hardware technologies, an old description in VHDL may be recycled and implemented differently.

VHDL [2, 3] allows several views on one component which yields to several ways of description. Here we want to concentrate on the behavioural description because it is similar to a TLA specification of this component. This type of VHDL-specifications describe the behaviour of a component as a reaction at the outputs to a change of the values at the inputs. The advantage of this type of comportmental description is that designs, complicated in terms of real structure, can be described with just a few lines of VHDL code. VHDL differentiates between sequential and concurrent statements used for the behavioural description. The control structure elements for sequential statements are known from basic programming languages. There are mechanisms like conditional execution, loops, subroutines, functions and procedures, assignment operations and so on.

Among the set of concurrent language elements there are signal assignments as well as assertion statements and processes.

With VHDL we will use the most common hardware description language to translate a high level system specification into. There are a lot of tools that support the design of hardware devices out of a behavioural description (e.g.: the synopsis compiler). Many approaches have been discussed in order trying to fix the semantic of VHDL to be able to verify a described system. Here in this article we want to use the standard VHDL like it was funded and standardized by IEEE because the desired system can be verified in TLA.

3. TRANSFORMATION OF SPECIFICATIONS

The main idea of our approach is to specify and verify a system in cTLA+ and afterwards translate the specification into VHDL. But in order to transform the protocol specification from TLA to VHDL it is necessary to identify similarities and identical mechanisms between both languages. The great advantage of this approach is that both languages describe state transition systems. With a TLA specification all possible trajectories of a system, so all reachable states will be described. A usual VHDL implementation of a system will implement one behaviour because it is described in a deterministic way. So in the best case however, it would implement one possible trajectory which was defined with the TLA specification. But to translate the ideas of TLA into VHDL we have to adopt both specifications in order to get a straight forward translation of the TLA specification.

3.1 Enhancing cTLA+ for translation into VHDL

In TLA all actions describing the next state relations are predicates over two states. Therefore all the parameters of an action are used to evaluate if the action is enabled or not. This defines the logic view of describing a system in which the quantifier "exists" can be used to declare that there will be a value for the action-parameter that the action can take place. But in real systems the parameters of processes are used to export stored or evaluated data to the rest of the system. This is by definition not modelled in a TLA specification. In roughly speaking in TLA only the control graph of a system is modelled which is obviously not independent from the data.

But to translate a TLA description into VHDL we need more information about the data. We have to know if it is generated inside an operation or if the operation just checks or stores incoming data-segments. This results in an information in which way the data is delivered. In cTLA+ it is not possible to specify any information about the data-flow. So we have developed a new dialect of cTLA+ which is named vTLA+ (VHDL near cTLA+). It is based on cTLA+ and adds the desired parts of specifications concerning the data-flow in a consistent way. In Fig. 2 the process *BufS* is shown with the new extension which is introduced with the keyword *VHDL*. Similar to the *ACTIONS* part of the description all the used actions are mentioned here a second time. But only the header of an action with its used parameters need to be specified, because we want to describe the direction of the data which is "flowing" through these parameters. Therefore we have to denote for each action parameter if it is incoming, outgoing or both in a single statement like it is shown in Fig. 2.

```
PROCESS BufS
BODY
   VARIABLES ...............
   INIT == .......
   ACTIONS
        NReq(m : message_typ) ==  .......
        PReq(f : frametyp) ==  .........
   VHDL                              -- New extension starts with VHDL
        NReq(m : message_typ)        -- For each action we have to list the parameters
           m        : in             -- and specify if it is incoming or what else
        PReq(f : frametyp)
           f.message : out
           f.ack     : in
END
```

Figure 2: vTLA+ description of BufS

3.2   Modelling processes in VHDL

To translate a specification from vTLA+ into VHDL we have to change the way of modelling systems in VHDL. Most of the designers are specifying one block of VHDL-code that has to implement the complete system. They are thinking in one finite state machine that will fulfil their properties. For complex systems especially for communication protocols this way is too difficult to handle, because modern protocols include thousands of different behaviours depending on the characteristics of the used base medium for communication.

In vTLA+ a protocol is described with several processes, where each process implements a certain part of the protocol and maybe fulfils one of the desired properties. To compose these processes to a complete system a system-module is used to construct all the joint actions that are required to set up the protocol. For each TLA-process we build a separate module in VHDL. This is suitable, because the processes in vTLA+ are completely independent state machines. Every process has its own local variables and communicates only via the action parameters during a joint action with the environment. So each process defines a local state graph that consists of the private variables of the process. Such a state graph is shown in Fig.3. It belongs to the process *BufS* we have defined in chapter 2.2. The two local variables *var.ub* and *var.lb* are defining the state space of this process.

BUFS

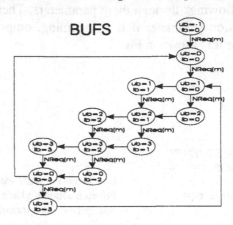

Figure 3: State transition graph of process BufS

What we have to do in VHDL is to implement the behaviour of the processes. This can be done by a straight forward translation of the vTLA+ description. The initial state description can be translated into the operations that will be executed when a RESET signal is received. The actions can be coded as simple operations, executed if a signal, which is named like the action itself, is received. So the intuitive translation of a vTLA+ process is really simple.

In TLA an action is a predicate about two states which can be true or in other words enabled or false otherwise. A joint action is a logical conjunction of the actions of several processes. So a joint action is only enabled iff all actions of the participating processes are enabled, too. But implementing these joint actions we can't start such an action without being sure that all actions of the participating processes could be executed. So we have to build a mechanism which will check if an action is executable or not, before we decide to start the joint action. Therefore we have to implement for each action a so-called enable-signal which is computed inside the process (it may depend on local variables of the process). These signals will be exported to the system-module which has to check all enable-signals for the used actions before the desired joint action can be started and all the participating actions of the processes can be executed. After processing the new state in all the processes the enable-signals have to be computed again. The system module in VHDL constructs the joint actions which are specified in TLA. It imports the submodules and couples the state graphs of these modules in a well defined way. This could lead to a very huge graph which can't be handled in just one module. Therefore the system module will just check the enable-signals of the participating process actions and will afterwards execute these actions. So the state graphs are coupled without constructing just one finite state machine. This leads to a very powerful but simple way of specifying large systems out of simple processes. Remembering these constraints the translation of a specification in vTLA+ into VHDL is straight forward.

4.   CONCLUSION

TLA or especially vTLA+ is a description language that uses logical formulas to describe the behaviours of a system. This can be done at the abstract level by specifying system properties or at the fine-grained level either. The advantage of a specification in TLA is the possibility of a formal proof. Modern communication protocols are highly complex, so well known mechanisms like model checking [5], on the fly verification [8] or symbolic evaluation aren't suitable anymore. So we would recommend using a formal proof which can be very easy in case of using formal methods for the system description. The formal verification of systems can be much more easier than testing the system in a simulation environment. Based on the compositionality of cTLA+ we can set up libraries of useful protocol elements which can be reused to support the design of new communication systems [13].

The upcoming new strategy of our approach is that the fine-grained system specifying the algorithm that implements the system can be translated into VHDL. This is done in a straight forward and tool supported manner. The strategy leads to a complete design-flow that supports the developer in order to produce a proven algorithm respectively a hardware-device which will implement the specified service.

A tool for translating TLA into VHDL or in other words a compiler will be built and the syntax of both languages can be described with a context free grammar. The complete compiler has to be developed so that we are able to translate all elements of the language vTLA+ into VHDL.

306

Testing the compiler includes a second part of research namely real time assumptions. For TLA exists an approach that includes real time specifications in the logic and allows reasoning about real time. We will adopt vTLA+ in the right way, so that we are able to guarantee the specified real time assumptions even in a hardware-device. This could lead to a very powerful development system and will generate a new high level description language.

References

[1] Alpern and Schneider. Defining liveness. Information Processing Letters, 1985.

[2] Armstrong, Gray. Structured Logic Design with VHDL. Prentice Hall, Englewood Cliffs

[3] IEEE Standard VHDL Language Reference Manual, New York, March 1988

[4] Clarke, Emerson and Sistla, Automatic verification of finite-state concurrent systems using temporal logic specifications. ACM Transactions of Programming Languages and Systems, April 1986.

[5] Clarke, Grumberg, Long, Model checking and abstraction. ACM Transactions of Programming Languages and Systems, 16(5):1512-1542, Sept. 1994

[6] Doeringer, Dykeman, Kaiserswerth et al., A survey of Light-Weight Transport Protocols for High-Speed Networks. IEEE Transactions on Communications, 1990

[7] Drögehorn. Ein Werkzeug zum formal basierten Entwurf von Hochleistungstransfer-protokollen. Diplomarbeit, Universität Dortmund, Informatik IV, 1996. In German.

[8] Fernandez and Mounier. "On the fly" verification of behavioural equivalences and preorders. Lecture Notes in Computer Science 575, 1991, Springer-Verlag.

[9] Geist. Eine Theorembeweiser-gestützte Entwicklungsumgebung für die halbautomatische Verifikation verteilter Systeme. Diplomarbeit, Universität Dortmund, Informatik IV, Dec. 1994. In German.

[10] Gerth, Peled, Vardi and Wolper. Simple On-the-fly Automatic Verification of Linear Temporal Logic. In P.Dembinski and M.Sredniawa. Protocol Specification, Testing, and Verification XV, p. 3-18, Warsaw, Poland, 1995. Chapman & Hall.

[11] Gibbs. Software's Chronic Crisis. Scientific American, 271(3):72-81, Sept. 1994.

[12] Herrmann and Krumm. Compositional Specification and Verification of High-Speed Transfer Protocols. In S.T.Voung and S.T.Chanson, Protocol Specification, Testing, and Verification XIV, IFIP Chapman & Hall.

[13] Herrmann and Krumm. Re-Useable Verification Elements for High-Speed Transfer Protocol Configurations. In P.Dembinski and M.Sredniawa, Protocol Specification, Testing, and Verification XV, Warshaw, 1995, Chapman & Hall.

[14] Holzman. Algorithms for Automated Protocol Verification. AT&T Technical Journal. p. 32-44, Jan. 1990.

[15] ISO. ESTELLE: A formal description technique based on an extended state transition model, International Standard ISO/IS 9074 edition, 1989

[16] ISO. LOTOS: Language for the temporal ordering specification of observational behaviour, International Standard ISO/IS 8807 edition, 1989

[17] ITU, SG X: Recommendation Z.100: CCITT Specification and Description Language SDL, 1993

[18] Lamport. The Temporal Logic of Actions. ACM Transactions on Programming Languages and Systems, May 1994